Linear Analysis

LINEAR ANALYSIS

An Introductory Course

Béla Bollobás
University of Cambridge

PUBLISHED BY THE PRESS SYNDICATE OF THE UNIVERSITY OF CAMBRIDGE
The Pitt Building, Trumpington Street, Cambridge CB2 1RP, United Kingdom

CAMBRIDGE UNIVERSITY PRESS
The Edinburgh Building, Cambridge CB2 2RU, UK http://www.cup.cam.ac.uk
40 West 20th Street, New York, NY 10011-4211, USA http://www.cup.org
10 Stamford Road, Oakleigh, Melbourne 3166, Australia

First edition © Cambridge University Press 1990
Second edition © Cambridge University Press 1999

First published 1990
Second edition 1999

Typeset in Times 10/13pt

A catalogue record for this book is available from the British Library

ISBN 0 521 65577 3 paperback

Transferred to digital printing 2004

To Márk

Qui cupit, capit omnia

CONTENTS

PREFACE

This book has grown out of the Linear Analysis course given in Cambridge on numerous occasions for the third-year undergraduates reading mathematics. It is intended to be a fairly concise, yet readable and down-to-earth, introduction to functional analysis, with plenty of challenging exercises. In common with many authors, I have tried to write the kind of book that I would have liked to have learned from as an undergraduate. I am convinced that functional analysis is a particularly beautiful and elegant area of mathematics, and I have tried to convey my enthusiasm to the reader.

In most universities, the courses covering the contents of this book are given under the heading of Functional Analysis; the name Linear Analysis has been chosen to emphasize that most of the material in on *linear* functional analysis. Functional Analysis, in its wide sense, includes partial differential equations, stochastic theory and non-commutative harmonic analysis, but its core is the study of normed spaces, together with linear functionals and operators on them. That core is the principal topic of this volume.

Functional analysis was born around the turn of the century, and within a few years, after an amazing burst of development, it was a well-established major branch of mathematics. The early growth of functional analysis was based on 19th century Italian function theory, and was given a great impetus by the birth of Lebesgue's theory of integration. The subject provided (and provides) a unifying framework for many areas: Fourier Analysis, Differential Equations, Integral Equations, Approximation Theory, Complex Function Theory, Analytic Number Theory, Measure Theory, Stochastic Theory, and so on.

From the very beginning, functional analysis was an international subject, with the major contributions coming from Germany, Hungary,
Poland, England and Russia: Fisher, Hahn, Hilbert, Minkowski and
Radon from Germany, Fejér, Haar, von Neumann, Frigyes Riesz and
Marcel Riesz from Hungary, Banach, Mazur, Orlicz, Schauder, Sierpiński
and Steinhaus from Poland, Hardy and Littlewood from England, Gelfand, Krein and Milman from Russia. The abstract theory of normed
spaces was developed in the 1920s by Banach and others, and was
presented as a fully fledged theory in Banach's epoch-making monograph,
published in 1932.

The subject of Banach's classic is at the heart of our course; this
material is supplemented with a body of other fundamental results and
some pointers to more recent developments.

The theory presented in this book is best considered as the natural
continuation of a sound basic course in general topology. The reader
would benefit from familiarity with measure theory, but he will not be at a
great disadvantage if his knowledge of measure theory is somewhat shaky
or even non-existent. However, in order to fully appreciate the power of
the results, and, even more, the power of the point of view, it is advisable
to look at the connections with integration theory, differential equations,
harmonic analysis, approximation theory, and so on.

Our aim is to give a fast introduction to the core of linear analysis, with
emphasis on the many beautiful general results concerning *abstract* spaces.
An important feature of the book is the large collection of exercises, many
of which are testing, and some of which are quite difficult. An exercise
which is marked with a plus is thought to be particularly difficult. (Need-
less to say, the reader may not always agree with this value judgement.)
Anyone willing to attempt a fair number of the exercises should obtain a
thorough grounding in linear analysis.

To help the reader, definitions are occasionally repeated, various basic
facts are recalled, and there are reminders of the notation in several
places.

The third-year course in Cambridge contains well over half of the contents of this book, but a lecturer wishing to go at a leisurely pace will find
enough material for two terms (or semesters). The exercises should cer-
tainly provide enough work for two busy terms.

There are many people who deserve my thanks in connection with this
book. Undergraduates over the years helped to shape the course;
numerous misprints were found by many undergraduates, including John
Longley, Gábor Megyesi, Anthony Quas, Alex Scott and Alan Stacey.

I am grateful to Dr Pete Casazza for his comments on the completed manuscript. Finally, I am greatly indebted to Dr Imre Leader for having suggested many improvements to the presentation.

Cambridge, May 1990   Béla Bollobás

For this second edition, I have taken the opportunity to correct a number of errors and oversights. I am especially grateful to R. B. Burckel for providing me with a list of errata.

B. B.

1. BASIC INEQUALITIES

The arsenal of an analyst is stocked with inequalities. In this chapter we present briefly some of the simplest and most useful of these. It is an indication of the size of the subject that, although our aims are very modest, this chapter is rather long.

Perhaps the most basic inequality in analysis concerns the arithmetic and geometric means; it is sometimes called the AM-GM inequality. The *arithmetic mean* of a sequence $a = (a_1, \ldots, a_n)$ of n reals is

$$A(a) = \frac{1}{n} \sum_{i=1}^{n} a_i;$$

if each a_i is non-negative then the *geometric mean* is

$$G(a) = \left(\prod_{i=1}^{n} a_i \right)^{1/n},$$

where the non-negative nth root is taken.

Theorem 1. The geometric mean of n non-negative reals does not exceed their arithmetic mean: if $a = (a_1, \ldots, a_n)$ then

$$G(a) \leqslant A(a). \tag{1}$$

Equality holds iff $a_1 = \cdots = a_n$.

Proof. This inequality has many simple proofs; the witty proof we shall present was given by Augustin-Louis Cauchy in his Cours d'Analyse (1821). (See Exercise 1 for another proof.) Let us note first that the theorem holds for $n = 2$. Indeed,

$$(a_1 - a_2)^2 = a_1^2 - 2a_1 a_2 + a_2^2 \geqslant 0;$$

so

$$(a_1 + a_2)^2 \geqslant 4a_1 a_2,$$

with equality iff $a_1 = a_2$.

Suppose now that the theorem holds for $n = m$. We shall show that it holds for $n = 2m$. Let $a_1, \ldots, a_m, b_1, \ldots, b_m$ be non-negative reals. Then

$$
\begin{aligned}
(a_1 \ldots a_m b_1 \ldots b_m)^{1/2m} &= \{(a_1 \ldots a_m)^{1/m}(b_1 \ldots b_m)^{1/m}\}^{1/2} \\
&\leqslant \tfrac{1}{2}\{(a_1 \ldots a_m)^{1/m} + (b_1 \ldots b_m)^{1/m}\} \\
&\leqslant \frac{1}{2}\left(\frac{a_1 + \ldots + a_m}{m} + \frac{b_1 + \ldots + b_m}{m}\right) \\
&= \frac{a_1 + \ldots + a_m + b_1 + \ldots + b_m}{2m}.
\end{aligned}
$$

If equality holds then, by the induction hypothesis, we have $a_1 = \cdots = a_m = b_1 = \cdots = b_m$. This implies that the theorem holds whenever n is a power of 2.

Finally, suppose n is an arbitrary integer. Let

$$n < 2^k = N \qquad \text{and} \qquad a = \frac{1}{n}\sum_{i=1}^{n} a_i.$$

Set $a_{n+1} = \cdots = a_N = a$. Then

$$\prod_{i=1}^{N} a_i = a^{N-n}\prod_{i=1}^{n} a_i \leqslant \left(\frac{1}{N}\sum_{i=1}^{N} a_i\right)^N = a^N;$$

so

$$\prod_{i=1}^{n} a_i \leqslant a^n,$$

with equality iff $a_1 = \cdots = a_N$, in other words iff $a_1 = \cdots = a_n$. $\square$

In 1906 Jensen obtained some considerable extensions of the AM-GM inequality. These extensions were based on the theory of convex functions, founded by Jensen himself.

A subset D of a real vector space is *convex* if every convex linear combination of a pair of points of D is in D, i.e. if $x, y \in D$ and $0 < t < 1$ imply that $tx + (1-t)y \in D$. Note that if D is convex, $x_1, \ldots, x_n \in D$, $t_1, \ldots, t_n > 0$ and $\sum_{i=1}^{n} t_i = 1$ then $\sum_{i=1}^{n} t_i x_i \in D$.

Indeed, assuming that a convex linear combination of $n-1$ points of D is in D, we find that

$$x_2' = \sum_{i=2}^{n} \frac{t_i}{1-t_1} x_i \in D$$

and so

$$\sum_{i=1}^{n} t_i x_i = t_1 x_1 + (1-t_1) x_2' \in D.$$

Given a convex subset D of a real vector space, a function $f: D \to \mathbb{R}$ is said to be *convex* if

$$f(tx + (1-t)y) \leq tf(x) + (1-t)f(y) \tag{2}$$

whenever $x, y \in D$ and $0 < t < 1$. We call f *strictly convex* if it is convex and, moreover, $f(tx + (1-t)y) = tf(x) + (1-t)f(y)$ and $0 < t < 1$ imply that $x = y$. Thus f is strictly convex if strict inequality holds in (2) whenever $x \neq y$ and $0 < t < 1$. A function f is *concave* if $-f$ is convex and it is *strictly concave* if $-f$ is strictly convex. Clearly, f is convex iff the set $\{(x, y) \in D \times \mathbb{R} : y \geq f(x)\}$ is convex.

Furthermore, a function $f: D \to \mathbb{R}$ is convex (concave, ...) iff its restriction to every interval $[a, b] = \{ta + (1-t)b : 0 \leq t \leq 1\}$ in D is convex (concave, ...). Rolle's theorem implies that if $f: (a, b) \to \mathbb{R}$ is differentiable then f is convex iff f' is increasing and f is concave iff f' is decreasing. In particular, if f is twice differentiable and $f'' \geq 0$ then f is convex, while if $f'' \leq 0$ then f is concave. Also, if $f'' > 0$ then f is strictly convex and if $f'' < 0$ then f is strictly concave.

The following simple result is often called Jensen's theorem; in spite of its straightforward proof, the result has a great many applications.

Theorem 2. Let $f: D \to \mathbb{R}$ be a concave function. Then

$$\sum_{i=1}^{n} t_i f(x_i) \leq f\left(\sum_{i=1}^{n} t_i x_i \right) \tag{3}$$

whenever $x_1, \ldots, x_n \in D$, $t_1, \ldots, t_n \in (0, 1)$ and $\sum_{i=1}^{n} t_i = 1$. Furthermore, if f is strictly concave then equality holds in (3) iff $x_1 = \cdots = x_n$.

Proof. Let us apply induction on n. As for $n = 1$ there is nothing to prove and for $n = 2$ the assertions are immediate from the definitions, let us assume that $n \geq 3$ and the assertions hold for smaller values of n.

Suppose first that f is concave, and let

$$x_1,\ldots,x_n \in D, \qquad t_1,\ldots,t_n \in (0,1) \qquad \text{with } \sum_{i=1}^{n} t_i = 1.$$

For $i = 2,\ldots,n$, set $t_i' = t_i/(1-t_1)$, so that $\sum_{i=2}^{n} t_i' = 1$. Then, by applying the induction hypothesis twice, first for $n-1$ and then for 2, we find that

$$\sum_{i=1}^{n} t_i f(x_i) = t_1 f(x_1) + (1-t_1) \sum_{i=2}^{n} t_i' f(x_i)$$

$$\leq t_1 f(x_1) + (1-t_1) f\left(\sum_{i=2}^{n} t_i' x_i\right)$$

$$\leq f\left(t_1 x_1 + (1-t_1) \sum_{i=2}^{n} t_i' x_i\right)$$

$$= f\left(\sum_{i=1}^{n} t_i x_i\right).$$

If f is strictly concave, $n \geq 3$ and not all x_i are equal then we may assume that not all of $x_2,\ldots,x_n$ are equal. But then

$$(1-t_1) \sum_{i=2}^{n} t_i' f(x_i) < (1-t_1) f\left(\sum_{i=2}^{n} t_i' x_i\right);$$

so the inequality in (3) is strict.   □

It is very easy to recover the AM-GM inequality from Jensen's theorem: $\log x$ is a strictly concave function from $(0,\infty)$ to $\mathbb{R}$, so for $a_1,\ldots,a_n > 0$ we have

$$\frac{1}{n} \sum_{i=1}^{n} \log a_i \leq \log \sum_{i=1}^{n} \frac{a_i}{n},$$

which is equivalent to (1). In fact, if $t_1,\ldots,t_n > 0$ and $\sum_{i=1}^{n} t_i = 1$ then

$$\sum_{i=1}^{n} t_i \log x_i \leq \log \sum_{i=1}^{n} t_i x_i, \tag{4}$$

with equality iff $x_1 = \cdots = x_n$, giving the following extension of Theorem 1.

Theorem 3. Let $a_1,\ldots,a_n \geq 0$ and $p_1,\ldots,p_n > 0$ with $\sum_{i=1}^{n} p_i = 1$. Then

$$\prod_{i=1}^{n} a_i^{p_i} \leq \sum_{i=1}^{n} p_i a_i, \tag{5}$$

with equality iff $a_1 = \cdots = a_n$.

Proof. The assertion is trivial if some a_i is 0; if each a_i is positive, the assertion follows from (4). $\square$

The two sides of (5) can be viewed as two different means of the sequence $a_1, \ldots, a_n$: the left-hand side is a generalized geometric mean and the right-hand side is a generalized arithmetic mean, with the various terms and factors taken with different weights. In fact, it is rather natural to define a further extension of these notions.

Let us fix $p_1, \ldots, p_n > 0$ with $\sum_{i=1}^{n} p_i = 1$: the p_i will play the role of *weights* or *probabilities*. Given a continuous and strictly monotonic function $\varphi : (0, \infty) \to \mathbb{R}$, the φ-*mean* of a sequence $a = (a_1, \ldots, a_n)$ $(a_i > 0)$ is defined as

$$M_\varphi(a) = \varphi^{-1}\left(\sum_{i=1}^{n} p_i \varphi(a_i)\right).$$

Note that M_φ need not be rearrangement invariant: for a permutation π the φ-mean of a sequence $a_1, \ldots, a_n$ need not equal the φ-mean of the sequence $a_{\pi(1)}, \ldots, a_{\pi(n)}$. Of course, if $p_1 = \cdots = p_n = 1/n$ then every φ-mean is rearrangement invariant.

It is clear that

$$\min_{1 \leq i \leq n} a_i \leq M_\varphi(a) \leq \max_{1 \leq i \leq n} a_i.$$

In particular, the mean of a constant sequence $(a_0, \ldots, a_0)$ is precisely a_0.

For which pairs φ and ψ are the means M_φ and M_ψ comparable? More precisely, for which pairs φ and ψ is it true that $M_\varphi(a) \leq M_\psi(a)$ for every sequence $a = (a_1, \ldots, a_n)$ $(a_i > 0)$? It may seem a little surprising that Jensen's theorem enables us to give an exact answer to these questions (see Exercise 31).

Theorem 4. Let $p_1, \ldots, p_n > 0$ be fixed weights with $\sum_{i=1}^{n} p_i = 1$ and let $\varphi, \psi : (0, \infty) \to \mathbb{R}$ be continuous and strictly monotone functions, such that $\varphi\psi^{-1}$ is concave if φ is increasing and convex if φ is decreasing. Then

$$M_\varphi(a) \leq M_\psi(a)$$

for every sequence $a = (a_1,\ldots,a_n)$ $(a_i > 0)$. If $\varphi\psi^{-1}$ is strictly concave (respectively, strictly convex) then equality holds iff $a_1 = \cdots = a_n$.

Proof. Suppose that φ is increasing and $\varphi\psi^{-1}$ is concave. Set $b_i = \psi(a_i)$ and note that, by Jensen's theorem,

$$M_\varphi(a) = \varphi^{-1}\left(\sum_{i=1}^n p_i\varphi(a_i)\right) = \varphi^{-1}\left(\sum_{i=1}^n p_i(\varphi\psi^{-1})(b_i)\right)$$

$$\leq \varphi^{-1}\left((\varphi\psi^{-1})\left\{\sum_{i=1}^n p_i b_i\right\}\right)$$

$$= \psi^{-1}\left(\sum_{i=1}^n p_i\psi(a_i)\right) = M_\psi(a).$$

If $\varphi\psi^{-1}$ is strictly concave and not all a_i are equal then the inequality above is strict since not all b_i are equal.

The case when φ is decreasing and $\varphi\psi^{-1}$ is convex is proved analogously. $\square$

When studying the various means of positive sequences, it is convenient to use the convention that a stands for a sequence $(a_1,\ldots,a_n)$, b for a sequence $(b_1,\ldots,b_n)$ and so on; furthermore,

$$a^{-1} = \frac{1}{a} = (a_1^{-1},\ldots,a_n^{-1}), \qquad a+x = (a_1+x,\ldots,a_n+x) \qquad (x \in \mathbb{R}^+),$$

$$ab = (a_1 b_1,\ldots,a_n b_n), \qquad abc = (a_1 b_1 c_1,\ldots,a_n b_n c_n),$$

and so on.

If $\varphi(t) = t^r$ $(-\infty < r < \infty, r \neq 0)$ then one usually writes M_r for M_φ. For $r > 0$ we define the mean M_r for all non-negative sequences: if $a = (a_1,\ldots,a_n)$ $(a_i \geq 0)$ then

$$M_r(a) = \left(\sum_{i=1}^n p_i a_i^r\right)^{1/r}.$$

Note that if $p_1 = \cdots = p_n = 1/n$ then M_1 is the usual *arithmetic mean* A, M_2 is the *quadratic mean* and M_{-1} is the *harmonic mean*. As an immediate consequence of Theorem 4, we shall see that M_r is a continuous monotone increasing function of r.

In fact, $M_r(a)$ has a natural extension from $(-\infty,0) \cup (0,\infty)$ to the whole of the extended real line $[-\infty,\infty]$ such that $M_r(a)$ is a continuous monotone increasing function. To be precise, put

$$M_\infty(a) = \max_{1 \le i \le n} a_i, \qquad M_{-\infty}(a) = \min_{1 \le i \le n} a_i, \qquad M_0(a) = \prod_{i=1}^{n} a_i^{p_i}.$$

Thus $M_0(a)$ is the weighted geometric mean of the a_i. It is easily checked that we have $M_r(a) = \{M_{-r}(a^{-1})\}^{-1}$ for all r $(-\infty \le r \le \infty)$.

Theorem 5. Let $a = (a_1, \ldots, a_n)$ be a sequence of positive numbers, not all equal. Then $M_r(a)$ is a continuous and strictly increasing function of r on the extended real line $-\infty \le r \le \infty$.

Proof. It is clear that $M_r(a)$ is continuous on $(-\infty, 0) \cup (0, \infty)$. To show that it is strictly increasing on this set, let us fix r and s, with $-\infty < r < s < \infty$, $r \ne 0$ and $s \ne 0$. If $0 < r$ then t^r is an increasing function of $t > 0$, and $t^{r/s}$ is a concave function, and if $r < 0$ then t^r is decreasing and $t^{r/s}$ is convex. Hence, by Theorem 4, we have $M_r(a) < M_s(a)$.

Let us write $A(a)$ and $G(a)$ for the *weighted* arithmetic and geometric means of $a = (a_1, \ldots, a_n)$, i.e. set

$$A(a) = M_1(a) = \sum_{i=1}^{n} p_i a_i \qquad \text{and} \qquad G(a) = M_0(a) = \prod_{i=1}^{n} a_i^{p_i}.$$

To complete the proof of the theorem, all we have to do is to show that

$$M_\infty(a) = \lim_{r \to \infty} M_r(a), \qquad M_{-\infty}(a) = \lim_{r \to -\infty} M_r(a), \qquad G(a) = \lim_{r \to 0} M_r(a).$$

The proofs of the first two assertions are straightforward. Indeed, let $1 \le m \le n$ be such that $a_m = M_\infty(a)$. Then for $r > 0$ we have

$$M_r(a) \ge (p_m a_m^r)^{1/r} = p_m^{1/r} a_m;$$

so $\liminf_{r \to \infty} M_r(a) \ge a_m = M_\infty(a)$. Since $M_r(a) \le M_\infty(a)$ for every r, we have $\lim_{r \to \infty} M_r(a) = M_\infty(a)$, as required. Also,

$$M_{-\infty}(a) = \{M_\infty(a^{-1})\}^{-1} = \{\lim_{r \to \infty} M_r(a^{-1})\}^{-1} = \lim_{r \to -\infty} M_r(a).$$

The final assertion, $G(a) = \lim_{r \to 0} M_r(a)$, requires a little care. In keeping with our conventions, for $-\infty < r < \infty$ $(r \ne 0)$ let us write $a^r = (a_1^r, \ldots, a_n^r)$. Then, clearly,

$$M_r(a) = A(a^r)^{1/r}.$$

Also, it is immediate that

$$\lim_{r \to 0} \frac{1}{r}(a_i^r - 1) = \frac{\partial}{\partial r} e^{r \log a_i} \bigg|_{r=0} = \log a_i$$

and so

$$\lim_{r \to 0} \frac{1}{r}\{A(a^r) - 1\} = \log G(a). \tag{6}$$

Since

$$\log t \leq t - 1$$

for every $t > 0$, if $r > 0$ then

$$\log G(a) = \frac{1}{r} \log G(a^r) \leq \frac{1}{r} \log A(a^r) \leq \frac{1}{r}\{A(a^r) - 1\}.$$

Letting $r \to 0$, we see from (6) that the right-hand side tends to $\log G(a)$ and so

$$\lim_{r \to 0+} \log M_r(a) = \lim_{r \to 0+} \frac{1}{r} \log A(a^r) = \log G(a),$$

implying

$$\lim_{r \to 0+} M_r(a) = G(a).$$

Finally,

$$\lim_{r \to 0-} M_r(a) = \lim_{r \to 0+} \{M_r(a^{-1})\}^{-1} = G(a^{-1})^{-1} = G(a). \qquad \square$$

The most frequently used inequalities in functional analysis are due to Hölder, Minkowski, Cauchy and Schwarz. Recall that a *hermitian form* on a complex vector space V is a function $\varphi : V \times V \to \mathbb{C}$ such that $\varphi(\lambda x + \mu y, z) = \lambda \varphi(x, z) + \mu \varphi(y, z)$ and $\varphi(y, x) = \overline{\varphi(x, y)}$ for all $x, y, z \in V$ and $\lambda, \mu \in \mathbb{C}$. (Thus $\varphi(x, \lambda y + \mu z) = \bar{\lambda} \varphi(x, y) + \bar{\mu} \varphi(x, z)$.) A hermitian form φ is said to be *positive* if $\varphi(x, x)$ is a positive real number for all $x \in V$ ($x \neq 0$).

Let $\varphi(\cdot, \cdot)$ be a positive hermitian form on a complex vector space V. Then, given $x, y \in V$, the value

$$\varphi(\lambda x + y, \lambda x + y) = |\lambda|^2 \varphi(x, x) + 2 \operatorname{Re}(\lambda \varphi(x, y)) + \varphi(y, y)$$

is real and non-negative for all $\lambda \in \mathbb{C}$. For $x \neq 0$, setting $\lambda = -\overline{\varphi(x, y)}/\varphi(x, x)$, we find that

$$|\varphi(x, y)|^2 \leq \varphi(x, x)\varphi(y, y)$$

and the same inequality holds, trivially, for $x = 0$ as well. This is the

Cauchy–Schwarz inequality. In particular, as

$$\varphi(x,y) = \sum_{i=1}^{n} x_i \bar{y}_i$$

is a positive hermitian form on $\mathbb{C}^n$,

$$\left| \sum_{i=1}^{n} x_i \bar{y}_i \right| \leqslant \left(\sum_{i=1}^{n} |x_i|^2 \right)^{1/2} \left(\sum_{i=1}^{n} |y_i|^2 \right)^{1/2},$$

and so

$$\left| \sum_{i=1}^{n} x_i y_i \right| \leqslant \left(\sum_{i=1}^{n} |x_i|^2 \right)^{1/2} \left(\sum_{i=1}^{n} |\bar{y}_i|^2 \right)^{1/2} = \left(\sum_{i=1}^{n} |x_i|^2 \right)^{1/2} \left(\sum_{i=1}^{n} |y_i|^2 \right)^{1/2}.$$

$$(7)$$

Our next aim is to prove an extension of (7), namely *Hölder's inequality.*

Theorem 6. Suppose

$$p, q > 1 \quad \text{and} \quad \frac{1}{p} + \frac{1}{q} = 1.$$

Then for complex numbers $a_1, \ldots, a_n, b_1, \ldots, b_n$ we have

$$\left| \sum_{k=1}^{n} a_k b_k \right| \leqslant \left(\sum_{k=1}^{n} |a_k|^p \right)^{1/p} \left(\sum_{k=1}^{n} |b_k|^q \right)^{1/q} \quad (8)$$

with equality iff all a_k are 0 or $|b_k|^q = t|a_k|^p$ and $a_k b_k = e^{i\theta}|a_k b_k|$ for all k and some t and θ.

Proof. Given non-negative reals a and b, set $x_1 = a^p$, $x_2 = b^q$, $p_1 = 1/p$ and $p_2 = 1/q$. Then, by Theorem 3,

$$ab = x_1^{p_1} x_2^{p_2} \leqslant p_1 x_1 + p_2 x_2 = \frac{a^p}{p} + \frac{b^q}{q}, \quad (9)$$

with equality iff $a^p = b^q$.

Hölder's inequality is a short step away from here. Indeed, if

$$\left(\sum_{k=1}^{n} |a_k| \right) \left(\sum_{k=1}^{n} |b_k| \right) \neq 0$$

then by homogeneity we may assume that

$$\sum_{k=1}^{n} |a_k|^p = \sum_{k=1}^{n} |b_k|^q = 1.$$

But then, by (9),

$$\left| \sum_{k=1}^{n} a_k b_k \right| \le \sum_{k=1}^{n} |a_k b_k| \le \sum_{k=1}^{n} \left(\frac{|a_k|^p}{p} + \frac{|b_k|^q}{q} \right) = \frac{1}{p} + \frac{1}{q} = 1.$$

Furthermore, if equality holds then

$$|a_k|^p = |b_k|^q \quad \text{and} \quad \left| \sum_{k=1}^{n} a_k b_k \right| = \sum_{k=1}^{n} |a_k b_k|,$$

implying $a_k b_k = e^{i\theta} |a_k b_k|$. Conversely, it is immediate that under these conditions we have equality in (8). $\qquad \square$

Note that if M_r denotes the rth mean with weights $p_i = n^{-1}$ ($i = 1, \ldots, n$) and for $a = (a_1, \ldots, a_n)$ and $b = (b_1, \ldots, b_n)$ we put $ab = (a_1 b_1, \ldots, a_n b_n)$, $|a| = (|a_1|, \ldots, |a_n|)$ and $|b| = (|b_1|, \ldots, |b_n|)$, then Hölder's inequality states that if $p^{-1} + q^{-1} = 1$ with $p, q > 1$, then

$$M_1(|ab|) \le M_p(|a|) M_q(|b|).$$

A minor change in the second half of the proof implies that (8) can be extended to an inequality concerning the means M_1, M_p and M_q with arbitrary weights (see Exercise 8).

The numbers p and q appearing in Hölder's inequality are said to be *conjugate exponents* (or *conjugate indices*). It is worth remembering that the condition $p^{-1} + q^{-1} = 1$ is the same as

$$(p-1)(q-1) = 1, \quad (p-1)q = p \quad \text{or} \quad (q-1)p = q.$$

Note that 2 is the only exponent which is its own conjugate. As we remarked earlier, the special case $p = q = 2$ of Hölder's inequality is called the Cauchy–Schwarz inequality.

In fact, one calls 1 and ∞ *conjugate exponents* as well. Hölder's inequality is essentially trivial for the pair $(1, \infty)$:

$$M_1(|ab|) \le M_1(|a|) M_\infty(|b|),$$

with equality iff there is a θ such that $|b_k| = M_\infty(|b|)$ and $a_k b_k = e^{i\theta} |a_k b_k|$ whenever $a_k \ne 0$.

The next result, *Minkowski's inequality*, is also of fundamental importance: in chapter 2 we shall use it to define the classical l_p spaces.

Theorem 7. Suppose $1 \le p < \infty$ and $a_1, \ldots, a_n, b_1, \ldots, b_n$ are complex numbers. Then

$$\left(\sum_{k=1}^{n} |a_k + b_k|^p \right)^{1/p} \le \left(\sum_{k=1}^{n} |a_k|^p \right)^{1/p} + \left(\sum_{k=1}^{n} |b_k|^p \right)^{1/p} \qquad (10)$$

with equality iff one of the following holds:
  (i) all a_k are 0;
  (ii) $b_k = ta_k$ for all k and some $t \geq 0$;
  (iii) $p = 1$ and, for each k, either $a_k = 0$ or $b_k = t_k a_k$ for some $t_k \geq 0$.

Proof. The assertion is obvious if $p = 1$ so let us suppose that $1 < p < \infty$, not all a_k are 0 and not all b_k are 0. Let q be the conjugate of p: $p^{-1} + q^{-1} = 1$. Note that

$$\sum_{k=1}^{n} |a_k + b_k|^p \leq \sum_{k=1}^{n} |a_k + b_k|^{p-1} |a_k| + \sum_{k=1}^{n} |a_k + b_k|^{p-1} |b_k|.$$

Applying (8), Hölder's inequality, to the two sums on the right-hand side with exponents q and p, we find that

$$\sum_{k=1}^{n} |a_k + b_k|^p$$

$$\leq \left(\sum_{k=1}^{n} |a_k + b_k|^{(p-1)q} \right)^{1/q} \left\{ \left(\sum_{k=1}^{n} |a_k|^p \right)^{1/p} + \left(\sum_{k=1}^{n} |b_k|^p \right)^{1/p} \right\}$$

$$= \left(\sum_{k=1}^{n} |a_k + b_k|^p \right)^{1/q} \left\{ \left(\sum_{k=1}^{n} |a_k|^p \right)^{1/p} + \left(\sum_{k=1}^{n} |b_k|^p \right)^{1/p} \right\}.$$

Dividing both sides by

$$\left(\sum_{k=1}^{n} |a_k + b_k|^p \right)^{1/q},$$

we obtain (8). The case of equality follows from that in Hölder's inequality. $\qquad\square$

Minkowski's inequality is also essentially trivial for $p = \infty$, i.e. for the M_∞ mean:

$$M_\infty(|a+b|) \leq M_\infty(|a|) + M_\infty(|b|),$$

with equality iff there is an index k such that $|a_k| = M_\infty(|a|)$, $|b_k| = M_\infty(|b|)$ and $|a_k + b_k| = |a_k| + |b_k|$.

The last two theorems are easily carried over from sequences to integrable functions, either by rewriting the proofs, almost word for word, or by approximating the functions by suitable step functions. Readers unfamiliar with Lebesgue measure will lose nothing if they take f and g to be piecewise continuous functions on $[0, 1]$.

Theorem 8. (Hölder's inequality for functions) Let p and q be conjugate indices and let f and g be measurable complex-valued functions on a measure space $(X, \mathcal{F}, \mu)$ such that $|f|^p$ and $|g|^q$ are integrable. Then fg is integrable and

$$\left| \int fg \, d\mu \right| \leq \left(\int |f|^p \, d\mu \right)^{1/p} \left(\int |g|^q \, d\mu \right)^{1/q}. \qquad \square$$

Theorem 9. (Minkowski's inequality for functions) Let $1 \leq p < \infty$ and let f and g be measurable complex-valued functions on a measure space $(X, \mathcal{F}, \mu)$ such that $|f|^p$ and $|g|^p$ are integrable. Then $|f+g|^p$ is integrable and

$$\left(\int |f+g|^p \, dx \right)^{1/p} \leq \left(\int |f|^p \, dx \right)^{1/p} + \left(\int |g|^p \, dx \right)^{1/p}. \qquad \square$$

Exercises

> All analysts spend half their time hunting through the literature for inequalities which they want to use but cannot prove.
>
> Harald Bohr

1. Let $A_n = \left\{ x = (x_i)_1^n : \sum_{i=1}^{n} x_i = n \text{ and } x_i \geq 0 \text{ for every } i \right\} \subset \mathbb{R}^n$.

   (i) Show that $g(x) = \prod_{i=1}^n x_i$ is bounded on A_n and attains its supremum at some point $z = (z_i)_1^n \in A_n$.

   (ii) Suppose that $x \in A$ and $x_1 = \min x_i < x_2 = \max x_i$. Set $y_1 = y_2 = \frac{1}{2}(x_1 + x_2)$ and $y_i = x_i$ for $3 \leq i \leq n$. Show that $y = (y_i)_1^n \in A_n$ and $g(y) > g(x)$. Deduce that $z_i = 1$ for all i.

   (iii) Deduce the AM-GM inequality.

2. Show that if $\psi : (a, b) \to (c, d)$ and $\varphi : (c, d) \to \mathbb{R}$ are convex functions and φ is increasing then $(\varphi \circ \psi)(x) = \varphi(\psi(x))$ is convex.

3. Suppose that $f : (a, b) \to (0, \infty)$ is such that $\log f$ is convex. Prove that f is convex.

4. Let $f : (a, b) \to (c, b)$ and $\varphi : (c, b) \to \mathbb{R}$ be such that φ and $\varphi^{-1} \circ f$ are convex. Show that f is convex.

5. Let $\{ f_\gamma : \gamma \in \Gamma \}$ be a family of convex functions on (a, b) such that $f(x) = \sup_{\gamma \in \Gamma} f_\gamma(x) < \infty$ for every $x \in (a, b)$. Show that $f(x)$ is also convex.

6. Suppose that $f : (0, 1) \to \mathbb{R}$ is an infinitely differentiable strictly convex function. Is it true that $f''(x) > 0$ for every $x \in (0, 1)$?

7. Let $p, q > 1$ be conjugate exponents. By considering the areas of the domains

$$D_1 = \{(x, y) : 0 \leqslant x \leqslant a \text{ and } 0 \leqslant y \leqslant x^{p-1}\}$$

and

$$D_2 = \{(x, y) : 0 \leqslant y \leqslant b \text{ and } 0 \leqslant x \leqslant y^{q-1}\},$$

prove inequality (8) and hence deduce Hölder's theorem. [Note that $(p-1)(q-1) = 1$, so if $y = x^{p-1}$ then $x = y^{q-1}$.]

8. Prove the following form of Hölder's theorem for the means

$$M_r(a) = M_r((a_i)_1^n) = \left(\sum_{i=1}^{n} p_i a_i^r \right)^{1/r} \qquad (r > 0).$$

If $p, q > 1$ are conjugate exponents, that is $p^{-1} + q^{-1} = 1$, and $(a_1, \ldots, a_n)$ and $(b_1, \ldots, b_n)$ are complex sequences then

$$\left| \sum_{i=1}^{n} p_i a_i b_i \right| \leqslant M_p(|a|) M_q(|b|),$$

with equality as in Theorem 6.

9. Deduce from the inequality in the previous exercise that if $a = (a_1, \ldots, a_n)$ is a positive sequence, that is, a sequence of positive reals, not all equal, then $M_r^r(a)$ is a strictly log-convex function of r, i.e. $r \log M_r(a)$ is a strictly convex function of r.

10. Show that for a fixed positive sequence a, $\log M_{1/r}(a)$ is a monotone decreasing convex function of $r > 0$, that is, if $r^{-1} = \frac{1}{2}(p^{-1} + q^{-1})$ then

$$M_r(a) \leqslant \{M_p(a) M_q(a)\}^{1/2}.$$

11. Show that $M_{1/r}(a)$ is a monotone decreasing convex function of r.

12. Show that if $0 < q < p < r$ then

$$\sum_{i=1}^{n} \mu_i a_i^p \leqslant \left(\sum_{i=1}^{n} \mu_i a_i^q \right)^{(r-p)/(r-q)} \left(\sum_{i=1}^{n} \mu_i a_i^r \right)^{(p-q)/(r-q)}$$

for all positive reals $a_1, a_2, \ldots, a_n, \mu_1, \mu_2, \ldots, \mu_n$.

13. Deduce from the previous exercise that if $0 < q < p < r$ and $f(x) > 0$ is continuous on (a, b) (or just measurable) then

$$\int_a^b f^p \, dx \leqslant \left(\int_a^b f^q \, dx \right)^{(r-p)/(r-q)} \left(\int_a^b f^r \, dx \right)^{(p-q)/(r-q)}.$$

14. Show that if $p, q, r > 0$ are such that $p^{-1} + q^{-1} = r^{-1}$ then

$$M_r(ab) \leqslant M_p(a) M_q(b)$$

for all positive sequences $a = (a_1, \ldots, a_n)$ and $b = (b_1, \ldots, b_n)$. Prove also that if $p, q, r, s > 0$ are such that $p^{-1} + q^{-1} + r^{-1} = s^{-1}$ then

$$M_s(abc) \leqslant M_p(a) M_q(b) M_r(c)$$

for all non-negative sequences a, b and c.

State and prove the analogous inequality for k sequences.

15. Let a and b be positive sequences. Show that if $r < 0 < s < t$ and $r^{-1} + s^{-1} = t^{-1}$ then

$$M_t(ab) \geqslant M_r(a) M_s(b).$$

Deduce that if $0 < r < 1$ then

$$M_r(a + b) \geqslant M_r(a) + M_r(b).$$

16. Let $\Phi_{p,q}(x, y) = x^{1/p} y^{1/q}$ $(x, y \geqslant 0)$. Show that $\Phi_{p,q}$ is concave iff $p, q \geqslant 1$ and $(p-1)(q-1) \geqslant 1$ and that it is convex iff $p, q \leqslant 1$ and $(p-1)(q-1) \geqslant 1$.

17. Deduce from the result in the previous exercise that $M_1(ab)$ and $M_p(a) M_q(b)$ are comparable if $(p-1)(q-1) \geqslant 1$. To be precise, if $p, q > 1$ and $(p-1)(q-1) \geqslant 1$ then

$$M_1(ab) \leqslant M_p(a) M_q(b)$$

and if $p, q < 1$ and $(p-1)(q-1) \geqslant 1$ then

$$M_1(a) \geqslant M_p(a) M_q(b).$$

18. $M_r(a + x) - x = M_r(a_1 + x, a_2 + x, \ldots, a_n + x) - x$ is a decreasing function of $x \geqslant 0$ and tends to $M_1(a)$ as $x \to \infty$. Show also that if $r \leqslant 1$ then $M_r(a + x) - x$ is an increasing function of $x \geqslant 0$ and it also tends to $M_1(a)$ as $x \to \infty$.

19. Prove that if $r > 1$, $a > 0$ and $a \neq 1$ then

$$a^r - 1 > r(a - 1)$$

and

$$a^{1/r} - 1 < \frac{a - 1}{r}.$$

Deduce that if $a, b > 0$ and $a \neq b$ then

$$rb^{r-1}(a-b) < a^r - b^r < ra^{r-1}(a-b)$$

if $r < 0$ or $r > 1$, and that the reverse inequalities hold if $0 < r < 1$.

20. Prove Chebyshev's inequality: if $r > 0$, $0 < a_1 \le a_2 \le \cdots \le a_n$ and $0 < b_1 \le b_2 \le \cdots \le b_n$ then

$$M_r(a)\,M_r(b) < M_r(ab)$$

unless all the a_i or all the b_i are equal. Prove also that the inequality is reversed if $a = (a_i)_1^n$ is monotone increasing and $b = (b_i)_1^n$ is monotone decreasing. [HINT: Note first that it suffices to prove the result for $r = 1$. For $r = 1$ the difference of the two sides is $\sum t_i \sum t_j a_j b_j - \sum t_i a_i \sum t_j b_j$.]

21. For $a_1, \ldots, a_n > 0$ $(a_1 \neq a_2)$ let π_k be the arithmetic mean of the $\binom{n}{k}$ products of the form $\prod_{j=1}^k a_{i_j}$ $(1 \le i_1 < \cdots < i_k \le n)$. Show that the sequence $\pi_0, \pi_1, \ldots, \pi_n$ is strictly log-concave:

$$\pi_{k-1}\pi_{k+1} < \pi_k^2$$

for all k $(1 \le k \le n-1)$. [HINT: Apply induction on n. For $n = 2$ the assertion is just the AM-GM inequality for two terms. Now let $n \ge 3$ and denote by π_r' the appropriate average for $a_1, \ldots, a_{n-1}$. Check that

$$\pi_r = \frac{n-r}{n}\pi_r' + \frac{r}{n}a_n\pi_{r-1}'$$

and deduce that for

$$n^2(\pi_{r-1}\pi_{r+1} - \pi_r^2) = A + Ba_n + Ca_n^2$$

we have

$$A < -\pi_r'^2, \quad B < 2\pi_{r-1}'\pi_r', \quad C < -\pi_{r-1}'^2.]$$

22. Deduce from the inequality in the previous exercise the following considerable extension of the AM-GM inequality, already known to Newton: if $a_1, \ldots, a_n$ are positive, $n \ge 2$ and $a_1 \neq a_2$ then

$$\pi_1 > \pi_2^{1/2} > \pi_3^{1/3} > \cdots > \pi_n^{1/n}.$$

23. Let $f: \mathbb{R}^+ \to \mathbb{R}$ be a strictly convex function with $f(0) \le 0$. Prove that if $a_1, \ldots, a_n \ge 0$ and at least two a_i are non-zero then

$$\sum_{i=1}^n f(a_i) < f\left(\sum_{i=1}^n a_i\right).$$

24. Let $f: (0, \infty) \to (0, \infty)$ be a monotone increasing function such that $f(x)/x^2$ is monotone decreasing. For $a, b > 0$ set

$$f(a, b) = b^2 f\left(\frac{a}{b}\right) \quad \text{and} \quad g(a, b) = a^2 \Big/ f\left(\frac{a}{b}\right).$$

Prove that if $a_i, b_i > 0$ then

$$\left(\sum_{i=1}^{n} a_i b_i\right)^2 \leqslant \left(\sum_{i=1}^{n} f(a_i, b_i)\right)\left(\sum_{i=1}^{n} g(a_i, b_i)\right) \leqslant \left(\sum_{i=1}^{n} a_i^2\right)\left(\sum_{i=1}^{n} b_i^2\right).$$

$$(*)$$

25. Prove that if $f, g : (0, \infty) \times (0, \infty) \to (0, \infty)$ satisfy $(*)$ for all $a_i, b_i > 0$ then they are of the form given in the previous exercise.

26. Show that

$$\left(\sum_{i=1}^{n} a_i b_i\right)^2 \leqslant \left(\sum_{i=1}^{n} a_i^2 + \sum_{i=1}^{n} b_i^2\right) \sum_{i=1}^{n} \frac{a_i^2 b_i^2}{a_i^2 + b_i^2} \leqslant \left(\sum_{i=1}^{n} a_i^2\right)\left(\sum_{i=1}^{n} b_i^2\right)$$

for all real a_i, b_i with $a_i^2 + b_i^2 > 0$.

27. Let $b_1, b_2, \ldots, b_n$ be a rearrangement of the positive numbers $a_1, a_2, \ldots, a_n$. Prove that

$$\sum_{i=1}^{n} \frac{a_i}{b_i} \geqslant n.$$

28. Let $f(x) \geqslant 0$ be a convex function. Prove that

$$\int_0^\infty f^2 \, dx \leqslant \tfrac{2}{3} \max f \int_0^\infty f \, dx.$$

Show that $\tfrac{2}{3}$ is best possible.

29. Let $f: [0, a] \to \mathbb{R}$ be a continuously differentiable function satisfying $f(0) = 0$. Prove the following inequality due to G. H. Hardy:

$$\int_0^a \left(\frac{f(x)}{2x}\right)^2 dx \leqslant \int_0^a \{f'(x)\}^2 \, dx.$$

[HINT: Note that

$$f'(x) = x^{1/2}\left(\frac{f(x)}{x^{1/2}}\right)' + \frac{f(x)}{2x}$$

and so

$$\{f'(x)\}^2 \geqslant \left(\frac{f(x)}{2x}\right)^2 + \left(\frac{f(x)}{x^{1/2}}\right)' \frac{f(x)}{x^{1/2}}.\Bigg]$$

30. Show that Theorem 4 characterizes comparable means, i.e. if $M_\varphi(a) \leqslant M_\psi(a)$ for all $a = (a_i)_1^n$ $(a_i > 0)$ then $\varphi\psi^{-1}$ is concave if φ is increasing and convex if φ is decreasing.

31. In a paper the authors claimed that if $x_k \geq 0$ for $k = 1, 2, \ldots, n$ then

$$x_n^n \geq x_1 (2x_2 - x_1)(3x_3 - 2x_2) \ldots (nx_n - (n - 1) x_{n-1}).$$

Show that this is indeed true if $2x_2 \geq x_1, 3x_3 \geq 2x_2, \ldots, nx_n \geq (n - 1) x_{n-1}$, and equality holds if and only if $x_1 = \ldots = x_n$. Show also that the inequality need not hold if any of the $n - 1$ inequalities $kx_k \geq (k - 1) x_{k-1}$ fails.

Notes

The foundation of the theory of convex functions is due to J. L. W. V. Jensen, *Sur les fonctions convexes et les inégalités entre les valeurs moyennes*, Acta Mathematica, **30** (1906), 175–93. Much of this chapter is based on the famous book of G. H. Hardy, J. E. Littlewood and G. Pólya, *Inequalities*, Cambridge University Press, First edition 1934, Second edition 1952, reprinted 1978, xii + 324 pp. This classic is still in print, and although its notation is slightly old-fashioned, it is well worth reading.

Other good books on inequalities are D. S. Mitrinovic, *Analytic Inequalities*, Springer-Verlag, Berlin and New York, 1970, xii + 400 pp., and A. W. Marshall and I. Olkin, *Inequalities: Theory of Majorization and Its Applications*, Academic Press, New York, 1979, xx + 569 pp.

2. NORMED SPACES AND BOUNDED LINEAR OPERATORS

In this long chapter we shall introduce the main objects studied in linear analysis: normed spaces and linear operators. Many of the normed spaces encountered in practice are spaces of functions (in particular, functions on $\mathbb{N}$, i.e. sequences), and the operators are often defined in terms of derivatives and integrals, but we shall concentrate on the notions defined in abstract terms.

As so often happens when starting a new area in mathematics, the ratio of theorems to definitions is rather low in this chapter. However, the reader familiar with elementary linear algebra and the rudiments of the theory of metric spaces is unlikely to find it heavy going because the concepts to be introduced here are only slight extensions of various concepts arising in those areas. Moreover, the relatively barren patch is rather small: as we shall see, even the basic definitions lead to fascinating questions.

A *normed space* is a pair $(V, \|\cdot\|)$, where V is a vector space over $\mathbb{R}$ or $\mathbb{C}$ and $\|\cdot\|$ is a function from V to $\mathbb{R}^+ = \{r \in \mathbb{R} : r \geq 0\}$ satisfying

  (i) $\|x\| = 0$ iff $x = 0$;
 (ii) $\|\lambda x\| = |\lambda| \|x\|$ for all $x \in V$ and scalar λ;
(iii) $\|x+y\| \leq \|x\| + \|y\|$ for all $x, y \in V$.

We call $\|x\|$ the *norm* of the vector x: it is the natural generalization of the length of a vector in the Euclidean spaces $\mathbb{R}^n$ or $\mathbb{C}^n$. Condition (iii) is the *triangle inequality*: in a triangle a side is no longer than the sum of the lengths of the other two sides.

In most cases the scalar field may be taken to be either $\mathbb{R}$ or $\mathbb{C}$, even when, for the sake of simplicity, we specify one or the other. If we want to emphasize that the ground field is $\mathbb{C}$, say, then we write 'complex normed space', 'complex l_p space', 'complex Banach space', etc.

Furthermore, unless there is some danger of confusion, we shall identify a normed space $X = (V, \|\cdot\|)$ with its underlying vector space V, and call the vectors in V the *points* or *vectors* of X. Thus $x \in X$ means that x is a point of X, i.e. a vector in V. We also say that $\|\cdot\|$ is a *norm* on X.

Every normed space is a metric space and so a topological space, and we shall often make use of some basic results of general topology. Although this book is aimed at the reader who has encountered metric spaces and topological spaces before, we shall review some of the basic concepts of general topology. A *metric space* is a pair (X, d), where X is a set and d is a function from $X \times X$ into $\mathbb{R}^+ = [0, \infty)$ such that (i) $d(x, y) = 0$ iff $x = y$, (ii) $d(x, y) = d(y, x)$ for all $x, y \in X$ and (iii) $d(x, z) \leq d(x, y) + d(y, z)$ for all $x, y, z \in X$. We call $d(x, y)$ the *distance* between x and y; the function d is a *metric* on X. Condition (iii) is again the *triangle inequality*.

A *topology* τ on a set X is a collection of subsets of X such that (i) $\varnothing \in \tau$ and $X \in \tau$, (ii) τ is closed under arbitrary unions: if $U_\gamma \in \tau$ for $\gamma \in \Gamma$ then $\bigcup_{\gamma \in \Gamma} U_\gamma \in \tau$, and (iii) τ is closed under finite intersection: if $U_1, \ldots, U_n \in \tau$ then $\bigcap_{i=1}^{n} U_i \in \tau$. The elements of the collection τ are said to be *open* (in the topology τ). A *topological space* is a pair (X, τ), where X is a set and τ is a topology on X. If it is clear that the topology we take is τ then we do not mention τ explicitly and we call X a topological space.

If Y is a subset (also called a subspace) of a topological space (X, τ) then $\{Y \cap U : U \in \tau\}$ is a topology on Y, called the *subspace topology* or the topology *induced* by τ. In most cases every subset Y is considered to be endowed with the subspace topology.

Given a topological space X, a set $N \subset X$ is said to be a *neighbourhood* of a point $x \in X$ if there is an open set U such that $x \in U \subset N$. A subset of X is *closed* if its complement is open. Since the intersection of a collection of closed sets is closed, every subset A of X is contained in a unique minimal closed set $\bar{A} = \{x \in X : \text{every neighbourhood of } x \text{ meets } A\}$, called the *closure* of A.

It is often convenient to specify a topology by giving a basis for it. Given a topological space (X, τ), a *basis* for τ is a collection σ of subsets of X such that $\sigma \subset \tau$ and every set in τ is a union of sets from σ. Clearly, if $\sigma \subset \mathcal{P}(X)$, i.e. σ is a family of subsets of X, then σ is a basis for a topology iff

(i) every point of X is in some element of σ;
(ii) if $B_1, B_2 \in \sigma$ then $B_1 \cap B_2$ is a union of some sets from σ.

A *neighbourhood base* at a point x_0 is a collection ν of neighbourhoods of x_0 such that every neighbourhood of x_0 contains a member of ν.

There are numerous ways of constructing new topological spaces from old ones; let us mention here the possibility of taking products, to be studied in some detail in Chapter 8. Let (X, σ) and (Y, τ) be topological spaces. The *product topology* on $X \times Y = \{(x, y) : x \in X, y \in Y\}$ is the topology with basis $\{U \times V : U \in \sigma, V \in \tau\}$. Thus a set $W \subset X \times Y$ is open iff for every $(x, y) \in W$ there are open sets $U \subset X$ and $V \subset Y$ such that $(x, y) \in U \times V \subset W$.

If d is a metric on X then the open balls

$$D(x, r) = \{y \in X : d(x, y) < r\} \qquad (x \in X, r > 0)$$

form a basis for a topology. This topology is said to be *defined* or *induced* by the metric d; we also call it the topology of the metric space. Not every topology is induced by a metric; for example,

$$\tau = \{U \subset \mathbb{R} : U = \varnothing \text{ or the complement } \mathbb{R} \backslash U \text{ of } U \text{ is countable}\}$$

is a topology on $\mathbb{R}$ and it is easily seen that it is not induced by any metric.

Given topological spaces (X_1, τ_1) and (X_2, τ_2), a map $f : X_1 \to X_2$ is said to be *continuous* if $f^{-1}(U) \in \tau_1$ for every $U \in \tau_2$, i.e. if the inverse image of every open set is open.

A bijection f from X_1 to X_2 such that both f and f^{-1} are continuous is said to be a *homeomorphism*; furthermore, (X_1, τ_1) and (X_2, τ_2) are said to be *homeomorphic* if there is a homeomorphism from X_1 to X_2.

A sequence $(x_n)_1^{\infty}$ in a topological space (X, τ) is said to be *convergent* to a point $x_0 \in X$, denoted $x_n \to x_0$ or $\lim_{n \to \infty} x_n = x_0$, if for every neighbourhood N of x_0 there is an n_0 such that $x_n \in N$ whenever $n \geq n_0$. Writing S for the subspace $\{n^{-1} : n = 1, 2, \ldots\} \cup \{0\}$ of $\mathbb{R}$ with the Euclidean topology, we see that $\lim_{n \to \infty} x_n = x_0$ iff the map $f : S \to X$, given by $f(n^{-1}) = x_n$ and $f(0) = x_0$ is continuous.

The topology of a metric space is determined by its convergent sequences. Indeed, a subset of a metric space is closed iff it contains the limits of its convergent sequences.

If σ and τ are topologies on a set X and $\sigma \subset \tau$ then σ is said to be *weaker* (or *coarser*) than τ, and τ is said to be *stronger* (or *finer*) than σ. Thus σ is weaker than τ iff the formal identity map $(X, \tau) \to (X, \sigma)$ is continuous.

The topological spaces occuring in linear analysis are almost always Hausdorff spaces, and we often consider compact Hausdorff spaces. A topology τ on a set X is a *Hausdorff topology* if for any two points $x, y \in X$ there are disjoint open sets U_x and U_y such that $x \in U_x$ and $y \in U_y$. A topological space (X, τ) is *compact* if every open cover has a finite subcover, i.e. if whenever $X = \bigcup_{\gamma \in \Gamma} U_\gamma$, where each U_γ is an open set, then $X = \bigcup_{\gamma \in F} U_\gamma$ for some finite subset F of Γ. A subset A of a topological space (X, τ) is said to be compact if the topology on A induced by τ is compact. Every closed subset of a compact space is compact, and in a compact Hausdorff space a set is compact iff it is closed.

It is immediate that if K is a compact space and $f : K \to \mathbb{R}$ is continuous then f is bounded and attains its supremum on K. Indeed, if we had $f(x) < s = \sup\{f(y) : y \in K\}$ for every $x \in K$ then

$$\bigcup_{r < s} \{x \in K : f(x) < r\}$$

would be an open cover of K without a finite subcover.

Every normed space X is a metric space with the induced metric $d(x, y) = \|x - y\|$. Conversely, given a metric d on a vector space X, setting $\|x\| = d(x, 0)$ defines a norm on X iff $d(x, y) = d(x + z, y + z)$ and $d(\lambda x, \lambda y) = |\lambda| d(x, y)$ for all $x, y, z \in X$ and scalar λ. The induced metric in turn, defines a topology on X, the *norm topology*. We shall always freely consider a normed space as a metric space with the induced metric and a topological space with the induced topology, and we shall use the corresponding terminology.

Let X be a normed space. By a *subspace* of X we mean a linear subspace Y of the underlying vector space, endowed with the norm on X (to be pedantic, with the restriction of the norm to Y). A subspace is *closed* if it is closed in the norm topology. Given a set $Z \subset X$, the subspace spanned by Z is

$$\operatorname{lin} Z = \left\{ \sum_{k=1}^{n} \lambda_k z_k : z_k \in Z, \lambda_k \text{ scalar}, n = 1, 2, \dots \right\}:$$

it is called the *linear span* of Z and it is the minimal subspace containing Z.

A normed space is *complete* if it is complete as a metric space, i.e. if every Cauchy sequence is convergent: if $(x_n)_1^\infty \subset X$ is such that $d(x_n, x_m) = \|x_n - x_m\| \to 0$ as $\min\{n, m\} \to \infty$ then $(x_n)_1^\infty$ converges to some point x_0 in X (i.e. $d(x_n, x_0) = \|x_n - x_0\| \to 0$). A complete normed space is called a *Banach space*. It is easily seen that a subset of a complete metric space is complete iff it is closed; thus a subspace of a Banach space is complete iff it is closed.

A metric space is *separable* if it contains a countable dense set, i.e. a countable set whose closure is the whole space. A normed space is *separable* if as a metric space it is separable. Most normed spaces we shall consider are separable.

Given a normed space X, the *unit sphere* of X is

$$S(X) = \{x \in X : \|x\| = 1\}$$

and the (*closed*) *unit ball* is

$$B(X) = \{x \in X : \|x\| \leqslant 1\}.$$

More generally, the *sphere of radius r about a point x_0* (or *centre x_0*) is

$$S_r(x_0) = S(x_0, r) = \{x \in X : \|x - x_0\| = r\}$$

and the (*closed*) *ball of radius r about x_0* (or *centre x_0*) is

$$B_r(x_0) = B(x_0, r) = \{x \in X : \|x - x_0\| \leqslant r\}.$$

Occasionally we shall need the *open ball of radius r and centre x_0*,

$$D_r(x_0) = D(x_0, r) = \{x \in X : \|x - x_0\| < r\}.$$

Note that the sets $x + tB(X)$ ($t > 0$) form a neighbourhood base at the point x.

The definition of the norm implies that $B_r(x_0)$ is closed, $D_r(x_0)$ is open, and for $r > 0$ the interior of $B_r(x_0)$ is $\text{Int}\, B_r(x_0) = D_r(x_0)$. The closure of $D_r(x_0)$ is $\overline{D_r(x_0)} = B_r(x_0)$. Furthermore, the boundary $\partial B_r(x_0)$ of $B_r(x_0)$ is the sphere $S_r(x_0)$.

The definition also implies that if X is a normed space then the map $X \times X \to X$ given by $(x, y) \mapsto x + y$ is uniformly continuous. Similarly, the map $\mathbb{R} \times X \to X$ (or $\mathbb{C} \times X \to X$) given by $(\lambda, x) \mapsto \lambda x$ is continuous. Note that the norm function $\|\cdot\|$ from X to $\mathbb{R}^+$ is also continuous.

Let us give a host of examples of normed spaces. Most of these are important spaces, while some others are presented only to illustrate the definitions. The vector spaces we take are vector spaces of sequences or functions with pointwise addition and multiplication: if $x = (x_i)_1^\infty$ and $y = (y_i)_1^\infty$ then $\lambda x + \mu y = (\lambda x_i + \mu y_i)_1^\infty$; if $f = f(t)$ and $g = g(t)$ are functions then $(\lambda f + \mu g)(t) = \lambda f(t) + \mu g(t)$.

In the examples below, and throughout the book, various sets are often assumed to be non-empty when the definitions would not make sense otherwise. Thus $S \neq \varnothing$ in (ii), $T \neq \varnothing$ in (iii), and so on.

Examples 1. (i) The n-dimensional *Euclidean space*: the vector space is $\mathbb{R}^n$ or $\mathbb{C}^n$ and the norm is

$$\|x\| = \left(\sum_{i=1}^{n} |x_i|^2 \right)^{1/2},$$

where $x = (x_1, \ldots, x_n)$. The former is a *real* Euclidean space, the latter is a *complex* one.

(ii) Let S be any set and let $\mathscr{F}_b(S)$ be the vector space of all bounded scalar-valued functions on S. For $f \in \mathscr{F}_b(S)$ let

$$\|f\| = \sup_{s \in S} |f(s)|.$$

This norm is the *uniform* or *supremum norm*.

(iii) Let L be a topological space, let $X = C(L)$ be the vector space of all bounded continuous functions on L and set, as in example (ii),

$$\|f\| = \sup_{t \in L} |f(t)|.$$

(iv) This is a special, but very important, case of the previous example. Let K be a compact Hausdorff space and let $C(K)$ be the space of continuous functions on K, with the supremum norm

$$\|f\| = \|f\|_\infty = \sup\{|f(x)| : x \in K\}.$$

Since K is compact, $|f(x)|$ is bounded on K and attains its supremum.

(v) Let X be $\mathbb{R}^n$ or $\mathbb{C}^n$ and set

$$\|x\|_1 = \sum_{i=1}^{n} |x_i|.$$

This is the space l_1^n; the norm is the l_1-*norm*.

Also,

$$\|x\|_\infty = \max_{1 \le i \le n} |x_i|$$

is a norm, the l_∞-norm; the space it gives is l_∞^n.

(vi) Let $1 \le p < \infty$. For $x = (x_1, \ldots, x_n) \in \mathbb{R}^n$ (or $\mathbb{C}^n$) put

$$\|x\|_p = \left(\sum_{k=1}^{n} |x_k|^p \right)^{1/p}.$$

This defines the space l_p^n (real or complex); the norm is the l_p-*norm*. The notation is consistent with that in example (v) for l_1^n and also, in a natural way, with that for l_∞^n (see Exercise 6). Note that l_2^n is exactly the n-dimensional Euclidean space.

(vii) Let X consist of all continuous real-valued functions $f(t)$ on $\mathbb{R}$ that vanish outside a finite interval, and put

$$\|f\|_1 = \int_{-\infty}^{\infty} |f(t)| \; dt.$$

(viii) Let X consist of all continuous complex-valued functions on $[0,1]$ and for $f \in X$ put

$$\|f\|_2 = \left(\int_0^1 |f(t)|^2 \; dt \right)^{1/2}.$$

(ix) For $1 \leqslant p < \infty$ the space l_p consists of all scalar sequences $x = (x_1, x_2, \ldots)$ for which

$$\left(\sum_{i=1}^{\infty} |x_i|^p \right)^{1/p} < \infty.$$

The norm of an element $x \in l_p$ is

$$\|x\|_p = \left(\sum_{i=1}^{\infty} |x_i|^p \right)^{1/p}.$$

The space l_∞ consists of all bounded scalar sequences with

$$\|x\|_\infty = \sup_{1 \leqslant i < \infty} |x_i|$$

and c_0 is the space of all scalar sequences tending to 0, with the same norm $\|x\|_\infty$. As remarked earlier, we have both real and complex forms of these spaces.

(x) For $1 \leqslant p < \infty$ the space $L_p(0,1)$ consists of those Lebesgue measurable functions on $[0,1]$ for which

$$\|f\|_p = \left(\int_0^1 |f(t)|^p \; dt \right)^{1/p} < \infty.$$

Note that $L_1(0,1)$ is precisely the space of integrable functions.

Strictly speaking, a point of $L_p(0,1)$ is an *equivalence class* of functions, two functions being equivalent if they agree almost everywhere. Putting it another way, f_1 and f_2 are equivalent (and so are considered to be identical) if $\|f_1 - f_2\|_p = 0$.

(xi) By analogy with the previous examples, the astute reader will guess that $L_\infty(0,1)$ consists of all essentially bounded Lebesgue measurable functions on $[0,1]$, i.e. those functions f for which

$$\|f\|_\infty = \text{ess sup}|f(t)| < \infty.$$

Recall that the *essential supremum* ess sup$|f(t)|$ is defined as

$$\inf\{\|f|S\|_\infty : S \subset [0,1] \text{ and } [0,1]\backslash S \text{ has measure } 0\}$$

$$= \inf\{a : \text{the set } \{t : |f(t)| > a\} \text{ has measure } 0\},$$

where $f|S$ is the *restriction of f to S* and so

$$\|f|S\|_\infty = \sup\{|f(s)| : s \in S\}.$$

Once again, strictly speaking $L_\infty(0,1)$ consists of *equivalence classes* of functions bounded on $[0,1]$.

(xii) $L_p(\mathbb{R})$, $L_\infty(\mathbb{R})$, $L_p(a,\infty)$, $L_\infty(a,\infty)$, etc., are defined similarly to the spaces in (x) and (xi). Of these, once again, the Hilbert spaces $L_2(\mathbb{R})$, $L_2(a,\infty)$, etc., are the most important.

(xiii) Let $C^{(n)}(0,1)$ consist of the functions $f(t)$ on $(0,1)$ having n continuous and bounded derivatives, with norm

$$\|f\| = \sup\left\{\sum_{k=0}^{n} |f^{(k)}(t)| : 0 < t < 1\right\}.$$

(xiv) Let X consist of all polynomials

$$f(t) = \sum_{k=0}^{n} c_k t^k$$

of degree at most n, with norm

$$\|f\| = \sum_{k=0}^{n} (k+1)|c_k|.$$

(xv) Let X be the space of bounded continuous functions on $(-1,1)$ which are differentiable at 0, with norm

$$|f'(0)| + \sup_{|t|<1} |f(t)|.$$

(xvi) Let X consist of all finite trigonometric polynomials

$$f(t) = \sum_{k=-n}^{n} c_k e^{ikt}, \qquad (n = 1, 2, \ldots)$$

with norm

$$\|f\| = \left(\sum_{k=-n}^{n} |c_k|^2\right)^{1/2}.$$

(xvii) Let V be a real vector space with basis (e_1, e_2), and for $x \in V$ define

$$\|x\| = \inf\left\{ |a| + |b| + |c| : a, b, c \in \mathbb{R} \text{ and } x = ae_1 + be_2 + \frac{c(e_1 + e_2)}{\sqrt{2}} \right\}.$$

(xviii) Let V be as in (xvii) and set

$$\|x\| = \inf\left\{ |a| + |b| + |c| + |d| : a, b, c, d \in \mathbb{R} \right.$$

$$\left. \text{and } x = ae_1 + be_2 + \frac{c(e_1 + e_2) + d(e_1 - e_2)}{\sqrt{2}} \right\}.$$

(xix) Let V be an n-dimensional real vector space and let $v_1, v_2, \ldots, v_m$ be vectors spanning V. For $x \in V$ define

$$\|x\| = \inf\left\{ \sum_{i=1}^{m} |c_i| : c_i \in \mathbb{R} \text{ and } x = \sum_{i=1}^{m} c_i v_i \right\}.$$

Let us prove that the examples above are indeed normed spaces, i.e. that the underlying spaces are vector spaces and the functions $\|\cdot\|$ satisfy conditions (i)–(iii). It is obvious that conditions (i) and (ii) are satisfied in each example so we have to check only (iii), the triangle inequality. Then it will also be clear that the underlying space is a vector space.

Furthermore, the triangle inequality is obvious in examples (ii)–(v), (vii), (xi) and (xii)–(xv). In the rest, with the exception of the last three examples, the triangle inequality is precisely Minkowski's inequality in one of its many guises. Thus Theorem 1.7 (Minkowski's inequality for sequences) is just the triangle inequality in l_p^n:

$$\|x + y\|_p \leq \|x\|_p + \|y\|_p \tag{1}$$

whenever $x, y \in l_p^n$; this clearly implies the analogous inequality in l_p; Theorem 1.9 (Minkowski's inequality for functions) is precisely the triangle inequality in $L_p(0, 1)$:

$$\|f + g\|_p \leq \|f\|_p + \|g\|_p \tag{2}$$

whenever $f, g \in L_p(0, 1)$, etc. In turn, the triangle inequality implies that the underlying spaces are indeed vector spaces.

Examples (xvii)–(xix) are rather similar, with (xix) being the most general case; we leave the proof to the reader (Exercise 7).   □

The spaces l_p ($1 \leq p \leq \infty$) are the simplest *classical sequence spaces*, and the spaces $C(K)$ and $L_p(0, 1)$ ($1 \leq p \leq \infty$) are the simplest *classical function spaces.* These spaces have been extensively studied for over eighty years: much is known about them but, in spite of all this attention, many important questions concerning them are waiting to be settled. In many ways, the most pleasant and most important of all infinite-dimensional Banach spaces is l_2, the space of square-summable sequences. This space l_2 is the canonical example of a *Hilbert space*. Similarly, of the finite-dimensional normed spaces, the space l_2^n is central: this is the *n-dimensional Euclidean space*.

The spaces in Examples (i)–(vi), (ix)–(xiv) and (xvii)–(xix) are complete, the others are incomplete. Some of these are easily seen, we shall see the others later.

Let us remark that if X is a normed space then its completion $\bar{X}$ as a metric space has a natural vector space structure and a natural norm. Thus every normed space is a dense subspace of a Banach space. We shall expand on this later in this chapter.

Having mentioned the more concise formulations of Minkowski's inequality (inequalities (1) and (2)), let us draw attention to the analogous formulations of Hölder's inequality. Suppose p and q are conjugate indices, with $p = 1$ and $q = \infty$ permitted. If $x = (x_k)_1^\infty \in l_p$ and $y = (y_k)_1^\infty \in l_q$ then

$$\sum_{k=1}^{\infty} |x_k y_k| \leqslant \|x\|_p \|y\|_q. \tag{3}$$

Similarly, if $f \in L_p(0, 1)$ and $g \in L_q(0, 1)$ then fg is integrable and

$$\|fg\|_1 = \int_0^1 |fg| \, \mathrm{d}x \leqslant \|f\|_p \|g\|_q. \tag{4}$$

It is easy to describe the general form of a norm on a vector space V in terms of its open unit ball. Let $X = (V, \|\cdot\|)$ be a normed space and let $D = \{x \in V : \|x\| < 1\}$ be the open unit ball. Clearly $\|\cdot\|$ is determined by D: if $x \neq 0$ then $\|x\| = \inf\{t : t > 0, x \in tD\}$.

The set D has the following properties:
(i) if $x, y \in D$ and $|\lambda| + |\mu| \leqslant 1$ then $\lambda x + \mu y \in D$;
(ii) if $x \in D$ then $x + \epsilon D \subset D$ for some $\epsilon = \epsilon(x) > 0$;
(iii) for $x \in V$ ($x \neq 0$) there are non-zero scalars λ and μ such that $\lambda x \in D$ and $\mu x \notin D$.

A set D satisfying (i) is said to be *absolutely convex*: this property is a consequence of the triangle inequality. Property (ii) follows from the fact that D is an *open* ball, while (iii) holds since $\|x\| < \infty$ for every x and $\|y\| = 0$ iff $y = 0$.

Conversely, if $D \subset V$ satisfies (i)–(iii) then

$$q(x) = \inf\{t : t > 0, x \in tD\}$$

defines a norm on V, and in this norm D is the open unit ball. The function $q(x)$ is the *Minkowski functional* determined by D. Note that in conditions (i)–(iii) we do not assume any topology on V.

When studying normed spaces, it is often useful to adopt a geometric point of view and examine the geometry, i.e. the 'shape', of the unit ball.

Much of the material presented in this book concerns linear functionals and linear operators. Let X and Y be normed spaces over the same ground field. A *linear operator from X to Y* is a linear map between the underlying vector spaces, i.e. a map $T : X \to Y$ such that

$$T(\lambda_1 x_1 + \lambda_2 x_2) = \lambda_1 T(x_1) + \lambda_2 T(x_2)$$

for all $x_1, x_2 \in X$ and scalars λ_1 and λ_2. The vector space of linear operators from X to Y is denoted by $\mathcal{L}(X, Y)$. The *image* of T is $\operatorname{Im} T = \{Tx : x \in X\}$ and the *kernel* of T is $\operatorname{Ker} T = \{x : Tx = 0\}$. Clearly $\operatorname{Ker} T$ is a subspace of X and $\operatorname{Im} T$ is a subspace of Y. Furthermore, T is a vector-space isomorphism iff $\operatorname{Ker} T = \{0\}$ and $\operatorname{Im} T = Y$.

A linear operator $T \in \mathcal{L}(X, Y)$ is *bounded* if there is an $N > 0$ such that

$$\|Tx\| \leq N\|x\| \qquad \text{for all } x \in X.$$

We shall write $\mathcal{B}(X, Y)$ for the set of bounded linear operators from X to Y; $\mathcal{B}(X) = \mathcal{B}(X, X)$ is the set of bounded linear operators on X. Clearly $\mathcal{B}(X, Y)$ is a vector space. The operators in $\mathcal{L}(X, Y)\backslash\mathcal{B}(X, Y)$ are said to be *unbounded*. A *linear functional on X* is a linear operator from X into the scalar field. We write X' for the space of linear functionals on X, and X^* for the vector space of bounded linear functionals on X. It is often convenient to use the *bracket notation* for the value of a functional on an element: for $x \in X'$ and $f \in X'$ we set $\langle f, x \rangle = \langle x, f \rangle = f(x)$. This is in keeping with the fact that the map $X \times X' \to \mathbb{R}$ (or $\mathbb{C}$) given by $(x, f) \mapsto f(x)$ is bilinear.

Theorem 2. Let X and Y be normed spaces and let $T : X \to Y$ be a linear operator. Then the following conditions are equivalent:
  (i) T is continuous (as a map of the topological space X into the topological space Y);
  (ii) T is continuous at some point $x_0 \in X$;
  (iii) T is bounded.

Proof. The implication (i) $\Rightarrow$ (ii) is trivial.

(ii) $\Rightarrow$ (iii). Suppose T is continuous at x_0. Since $Tx_0 + B(Y)$ is a neighbourhood of $T(x_0)$, there is a $\delta > 0$ such that

$$x = x_0 + y \in x_0 + \delta B(X) \Rightarrow Tx = Tx_0 + Ty \in Tx_0 + B(Y).$$

Hence $\|y\| \leq \delta$ implies $\|Ty\| \leq 1$ and so $\|Tz\| \leq \delta^{-1}\|z\|$ for all z.

(iii) $\Rightarrow$ (i). Suppose $\|Tx\| \leq N\|x\|$ for all $x \in X$. Then if $\|x - y\| < \epsilon/N$ we have $\|Tx - Ty\| < \epsilon$.　　　　　　　□

Two normed spaces X and Y are said to be *isomorphic* if there is a linear map $T: X \to Y$ which is a topological isomorphism (i.e. a homeomorphism). We call X and Y *isometrically isomorphic* if there is a linear isometry from X to Y, i.e. if there is a bijective operator $T \in \mathcal{B}(X, Y)$ such that $T^{-1} \in \mathcal{B}(Y, X)$ and $\|Tx\| = \|x\|$ for all $x \in X$. Two norms, $\|\cdot\|_1$ and $\|\cdot\|_2$, on the same vector space V are said to be *equivalent* if they induce the same topology on V, i.e. if the formal identity map from $X_1 = (V, \|\cdot\|_1)$ to $X_2 = (V, \|\cdot\|_2)$ is a topological isomorphism. As an immediate consequence of Theorem 2, we see that if a linear map is a topological isomorphism then both the map and its inverse are bounded.

Corollary 3. Let X and Y be normed spaces and let $T \in \mathcal{L}(X, Y)$. Then T is a topological isomorphism iff $T \in \mathcal{B}(X, Y)$ and $T^{-1} \in \mathcal{B}(Y, X)$.

Two norms, $\|\cdot\|_1$ and $\|\cdot\|_2$, on the same vector space V are equivalent iff there are constants $c, d > 0$ such that

$$c\|x\|_1 \leq \|x\|_2 \leq d\|x\|_1$$

for all $x \in V$.　　　　　　　□

It is immediate that equivalence of norms is indeed an equivalence relation, and Corollary 3 implies that if a normed space is complete then it is also complete in every equivalent norm.

The equivalence of norms has an intuitive geometrical interpretation in terms of the (open or closed) unit balls of the normed spaces. Let $\|\cdot\|_1$ and $\|\cdot\|_2$ be norms on V, with closed unit balls B_1 and B_2. Then $\|\cdot\|_1$ and $\|\cdot\|_2$ are equivalent if and only if $B_2 \subset cB_1$ and $B_1 \subset dB_2$ for some constants c and d. Indeed, these relations hold if and only if $\|x\|_1 \leq c\|x\|_2$ and $\|x\|_2 \leq d\|x\|_1$ for all $x \in V$.

It is clear that $\mathcal{B}(X, Y)$ and X^* are vector spaces: $\mathcal{B}(X, Y)$ is a subspace of $\mathcal{L}(X, Y)$ and X^* is a subspace of $\mathcal{L}(X, \mathbb{R})$. In fact, they are also normed spaces with a natural norm. The *operator norm* or simply *norm* on $\mathcal{B}(X, Y)$ is given by

$$\|T\| = \inf\{N > 0: \|Tx\| \leqslant N\|x\| \text{ for all } x \in X\} = \sup\{\|Tx\| : \|x\| \leqslant 1\}.$$

Although this gives us the definition of the norm of a bounded linear functional as well, let us spell it out: the *norm* of $f \in X^*$ is

$$\|f\| = \inf\{N > 0: \ |f(x)| \leqslant N\|x\| \quad \text{for} \quad \text{all} \quad x \in X\} = \sup\{|f(x)|: \|x\| \leqslant 1\}.$$

Note that in these definitions the infimum is attained so

$$\|Tx\| \leqslant \|T\|\|x\| \quad \text{and} \quad |f(x)| \leqslant \|f\|\|x\| \quad \text{for all } x.$$

The terminology is justified by the following simple result.

Theorem 4. The function $\|\cdot\|$, defined above, is a norm on $\mathcal{B}(X, Y)$. If Y is complete then so is $\mathcal{B}(X, Y)$. In particular, X^* is a complete normed space.

Proof. Only the second assertion needs proof. Let $(T_n)_1^\infty$ be a Cauchy sequence in $\mathcal{B}(X, Y)$. Then $(T_n x)_1^\infty$ is a Cauchy sequence in Y for every $x \in X$ and so there is a unique $y \in Y$ such that $T_n x \to y$. Set $Tx = y$.

To complete the proof, all we have to check is that $T \in \mathcal{B}(X, Y)$ and $T_n \to T$.

Given $x_1, x_2 \in X$ and scalars λ_1 and λ_2, we have

$$\begin{aligned}
T(\lambda_1 x_1 + \lambda_2 x_2) &= \lim_{n \to \infty} T_n(\lambda_1 x_1 + \lambda_2 x_2) \\
&= \lim_{n \to \infty}\{\lambda_1 T_n x_1 + \lambda_2 T_n x_2\} \\
&= \lambda_1 \lim_{n \to \infty} T_n x_1 + \lambda_2 \lim_{n \to \infty} T_n x_2 \\
&= \lambda_1 T x_1 + \lambda_2 T x_2,
\end{aligned}$$

and so $T \in \mathcal{L}(X, Y)$.

Furthermore, given $\epsilon > 0$, there is an n_0 such that $\|T_n - T_m\| < \epsilon$ if $n, m \geqslant n_0$. Then for $x \in X$ and $m \geqslant n_0$ we have

$$\begin{aligned}
\|Tx - T_m x\| &= \|\lim_{n \to \infty} T_n x - T_m x\| \\
&= \|\lim_{n \to \infty} (T_n - T_m)x\| \\
&= \lim_{n \to \infty} \|(T_n - T_m)x\| \\
&\leqslant \limsup_{n \to \infty} \|T_n - T_m\|\|x\| \leqslant \epsilon\|x\|.
\end{aligned}$$

Therefore

$$\|Tx\| \leq \epsilon\|x\| + \|T_m x\| \leq (\epsilon + \|T_m\|)\|x\|,$$

and so $T \in \mathcal{B}(X, Y)$. Finally, from the same inequality we find that $\|T - T_m\| \leq \epsilon$.   □

If we extend the operator norm to the whole of $\mathcal{L}(X, Y)$ by putting $\|T\| = \sup\{\|Tx\| : \|x\| \leq 1\} = \infty$ for an unbounded operator, then $\mathcal{B}(X, Y)$ consists of the operators in $\mathcal{L}(X, Y)$ having finite norm.

The Banach space X^* is called the *dual* of X. For $T \in \mathcal{B}(X, Y)$ and $g \in Y^*$ define a function $T^*g : X \to \mathbb{R}$ (or $\mathbb{C}$) by

$$(T^*g)(x) = g(Tx).$$

Then T^*g is a linear functional on X and $T^* : Y^* \to X^*$ is easily seen to be a linear map. Furthermore, T^* is not only in $\mathcal{L}(Y^*, X^*)$ but is, in fact, a bounded linear operator since

$$|(T^*g)(x)| = |g(Tx)| \leq \|g\|\|Tx\| \leq \|g\|\|T\|\|x\|,$$

so that

$$\|T^*g\| \leq \|T\|\,\|g\| \quad \text{and} \quad \|T^*\| \leq \|T\|.$$

The operator T^* is the *adjoint* of T. In fact, as we shall see later (Theorem 3.9), $\|T^*\| = \|T\|$.

The definition of the adjoint looks even more natural in the bracket notation: T^*g is the linear functional satisfying

$$\langle T^*g, x \rangle = \langle g, Tx \rangle.$$

Given maps $T : X \to Y$ and $S : Y \to Z$, we can compose them: the map $ST : X \to Z$ is given by $(ST)(x) = (S \circ T)(x) = S(T(x))$. If S and T are linear then so is ST; if they are bounded linear operators then so is ST.

Theorem 5. Let X, Y and Z be normed spaces and let $T \in \mathcal{B}(X, Y)$, $S \in \mathcal{B}(Y, Z)$. Then $ST : X \to Z$ is bounded linear operator and

$$\|ST\| \leq \|S\|\|T\|.$$

In particular, $\mathcal{B}(X) = \mathcal{B}(X, X)$ is closed under multiplication (i.e. under the composition of operators).

Proof. Clearly

$$\|(ST)(x)\| = \|S(Tx)\| \leq \|S\|\|Tx\| \leq \|S\|\|T\|\|x\|. \qquad □$$

The last result shows that if X is a Banach space then $\mathcal{B}(X)$ is a *unital Banach algebra*: it is an algebra with a unit (identity element) which is also a Banach space, such that the identity has norm 1 and the norm of the product of two elements is at most the product of their norms.

Examples 6. (i) Define $T : l_2^n \to l_2^m$ by putting

$$Tx = (x_1, \ldots, x_l, 0, \ldots, 0), \qquad \text{where } l = \min\{n, m\}.$$

Then $T \in \mathcal{B}(l_2^n, l_2^m)$ and $\|T\| = 1$.

(ii) If A is any endomorphism of $\mathbb{R}^n$ then $A \in \mathcal{B}(l_p^n, l_q^n)$ for all p and q $(1 \leqslant p, q \leqslant \infty)$.

(iii) Define $S \in \mathcal{B}(l_p)$ by

$$Sx = (0, x_1, x_2, \ldots).$$

This is the *right shift* operator.

The *left shift* $T \in \mathcal{B}(l_p)$ is defined by

$$Tx = (x_2, x_3, \ldots).$$

Clearly S is an injection but not a surjection, T is a surjection but not an injection, $\|S\| = \|T\| = 1$, and $TS = I$ (the identity operator) but $ST \neq I$: $\operatorname{Ker} ST = \{(x_1, 0, \ldots) : x_1 \in \mathbb{R}\}$.

(iv) Let $C^{(k)} = C^{(k)}(0, 1)$ be the normed space of functions on $(0, 1)$ with k continuous and bounded derivatives, as in Examples 1 (xiii). Let $D : C^{(n)} \to C^{(n-1)}$ be the differentiation operator: $Df = f'$. Then $\|D\| = 1$.

(v) Let $T : C^{(n)} \to C^{(0)}$ be the formal identity map $Tf = f$. Then $\|T\| = 1$.

(vi) In order to define linear functionals on function spaces, let us introduce the following notation. Given sequences $x = (x_k)_1^\infty$ and $y = (y_k)_1^\infty$, let $(x, y) = \sum_{k=1}^\infty x_k y_k$ and let $x \cdot y$ be the sequence $(x_k y_k)_1^\infty$. Thus if $x \cdot y \in l_1$ then (x, y) is well defined.

Inequality (3), i.e. Hölder's inequality for sequences, can now be restated once again in the following concise form: if p and q are conjugate indices, with $1 \leqslant p, q \leqslant \infty$, $x \in l_p$, and $y \in l_q$, then $x \cdot y \in l_1$ and

$$\|x \cdot y\|_1 \leqslant \|x\|_p \|y\|_q. \tag{5}$$

Inequality (5) implies that for $y \in l_q$ the function $f_y(x) = (x, y)$ is a bounded linear functional on l_p and $\|f_y\| \leqslant \|y\|_q$. In fact, $\|f_y\| = \|y\|_q$, and for $1 \leqslant p < \infty$ the correspondence $y \mapsto f_y$ is a linear isometry which identifies l_q with l_p^* (see Exercise 1 of the next chapter). The dual of l_∞

contains l_1 as a rather small subspace (l_∞ and l_∞^* are not separable but l_1 is); however, l_1 is the dual of c_0, the closed subspace of l_∞ consisting of sequences tending to 0 (again, see Exercise 1 of the next chapter).

(vii) Let p and q be conjugate indices and $g \in L_q(0,1)$. Define, for $f \in L_p(0,1)$,

$$\varphi_g(f) = \int_0^1 fg \, dx.$$

Then, by Hölder's inequality (4), φ_g is a bounded linear functional on $L_p(0,1)$ and $\|\varphi_g\| \leq \|g\|_q$. In fact, $\|\varphi_g\| = \|g\|_q$, and for $1 \leq p < \infty$ the correspondence $g \mapsto \varphi_g$ is a linear isometry which identifies $L_q(0,1)$ with $L_p(0,1)^*$.

(viii) Let $C[0,1]$ be the space of continuous functions with the supremum norm: $\|f\| = \|f\|_\infty = \max\{|f(t)| : 0 \leq t \leq 1\}$. Let $0 \leq t_0 \leq 1$ and define $\delta_{t_0} : C[0,1] \to \mathbb{R}$ by $\delta_{t_0}(f) = f(t_0)$. Then δ_{t_0} is a bounded linear functional of norm 1.

(ix) Let X be the subspace of $C[0,1]$ consisting of differentiable functions with continuous derivative. Then $D : X \to C[0,1]$, defined by $Df = f'$, is an unbounded linear operator.

(x) Let V be the vector space of all scalar sequences $x = (x_k)_1^\infty$ with *finitely many non-zero terms*. Set $X_p = (V, \|\cdot\|_p)$, where

$$\|x\|_p = \left(\sum_{k=1}^\infty |x_k|^p \right)^{1/p} \qquad (1 \leq p < \infty)$$

and

$$\|x\|_\infty = \max\{|x_k| : 1 \leq k < \infty\}.$$

For $1 \leq r, s \leq \infty$ let $T_{r,s} : X_r \to X_s$ be the formal identity map: $T_{r,s}x = x$. Then $T_{r,s}$ is a linear operator for all r and s. If $1 \leq r \leq s \leq \infty$ then $T_{r,s}$ is bounded: in fact, $\|T_{r,s}\| = 1$, but if $r > s$ then $T_{r,s}$ is unbounded (see Exercise 27). □

As promised earlier in the chapter, let us say a few words about completions. Every metric space has a unique completion, and if the metric is induced by a norm then this completion is a normed space. Before examining the completion of a normed space, let us review briefly the basic facts about the completion of a metric space.

A metric space $\tilde{X}$ is said to be the *completion* of a metric space X if $\tilde{X}$ is complete and X is a dense subset of $\tilde{X}$. (A subset A of a topological space T is *dense* if its closure is the whole of T.) The completion is unique in the sense that if X is a dense subset of the complete metric spaces Y and Z then there is a unique continuous map $\varphi : Y \to Z$ whose restriction to X is the identity, and this map is an isometry onto Z. This is easily seen since if $y \in Y$ then y is the limit (in Y) of a sequence $(x_n)_1^\infty \subset X$. Then $(x_n)_1^\infty$ is a Cauchy sequence in Z and so it has a limit $z \in Z$. Since φ is required to be continuous, we must have $\varphi(y) = z$ and so φ, if it exists, is unique. On the other hand, if we define a map $\varphi : Y \to Z$ by setting $\varphi(y) = z$, then this map is clearly an isometry from Y onto Z.

The completion of a metric space is defined by taking equivalence classes of Cauchy sequences. To be precise, given a metric space X, let $[X]$ be the set of Cauchy sequences of points of X:

$$[X] = \{(x_n)_1^\infty : x_n \in X, \lim_{n \to \infty} \sup_{m \geq n} d(x_n, x_m) = 0\}.$$

Define an equivalence relation $\sim$ on $[X]$ by setting $(x_n)_1^\infty \sim (y_n)_1^\infty$ if $d(x_n, y_n) \to 0$, and let $\tilde{X} = [X]/\sim$ be the collection of equivalence classes with respect to $\sim$. For $\tilde{x}, \tilde{y} \in \tilde{X}$ set

$$d(\tilde{x}, \tilde{y}) = \lim_{n \to \infty} d(x_n, y_n),$$

where $(x_n)_1^\infty \in \tilde{x}$ and $(y_n)_1^\infty \in \tilde{y}$. It is immediate that $d : \tilde{X} \times \tilde{X} \to [0, \infty)$ is well defined, i.e. that $\lim_{n \to \infty} d(x_n, y_n)$ is independent of the representatives chosen. Furthermore, if $(x_n)_1^\infty \in \tilde{x}$, $(y_n)_1^\infty \in \tilde{y}$ and $(z_n)_1^\infty \in \tilde{z}$ then

$$d(\tilde{x}, \tilde{z}) = \lim_{n \to \infty} d(x_n, z_n) \leq \lim_{n \to \infty} \sup \{d(x_n, y_n) + d(y_n, z_n)\}$$

$$= \lim_{n \to \infty} d(x_n, y_n) + \lim_{n \to \infty} d(y_n, z_n)$$

$$= d(\tilde{x}, \tilde{y}) + d(\tilde{y}, \tilde{z}).$$

Finally, $d(\tilde{x}, \tilde{x}) = 0$, and if $d(\tilde{x}, \tilde{y}) = 0$ then $(x_n)_1^\infty \sim (y_n)_1^\infty$, so that $\tilde{x} = \tilde{y}$.

The metric space $(\tilde{X}, d)$ is complete. Indeed, let $(\tilde{x}^{(k)})$ be a Cauchy sequence in $\tilde{X}$ and let $(x_n^{(k)})_{n=1}^\infty \in \tilde{x}^{(k)}$ $(k = 1, 2, \ldots)$. Let n_k be such that

$$d(x_n^{(k)}, x_m^{(k)}) < 2^{-k} \qquad \text{if } n, m \geq n_k.$$

Set $x_k = x_{n_k}^{(k)}$. It is easily checked that $(x_k)_1^\infty \in [X]$ and that $\tilde{x}^{(k)}$ tends to the equivalence class of the sequence $(x_k)_1^\infty$.

Writing $[x]$ for the equivalence class of the constant sequence $x, x, \ldots$, we find that the map $X \to \tilde{X}$, given by $x \mapsto [x]$, is an isometry. Thus, with a slight abuse of terminology, X can be considered to be a subset of $\tilde{X}$. Clearly, X is a dense subset of $\tilde{X}$, since if $(x_n)_1^\infty$ is a representative of $\tilde{x} \in \tilde{X}$ and $d(x_n, x_m) < \epsilon$ whenever $n, m \geq n_0$ then $d(\tilde{x}, x_{n_0}) = d(\tilde{x}, [x_{n_0}]) \leq \epsilon$.

If X is not only a metric space but also a normed space then its completion has a natural Banach-space structure.

Theorem 7. For every normed space X there is a Banach space $\tilde{X}$ such that X is dense in $\tilde{X}$. This space $\tilde{X}$ is unique in the sense that if X is a dense subspace of a Banach space $\overset{=}{X}$ then there is a unique continuous map $\varphi : \tilde{X} \to \overset{=}{X}$ such that the restriction of φ to X is the identity; furthermore, this map φ is a linear isometry from $\tilde{X}$ to $\overset{=}{X}$.

Proof. Considering X as a metric space, with the metric induced by the norm, i.e. with $d(x, y) = \|x - y\|$, let $\tilde{X}$ be its completion. Given $x, y \in \tilde{X}$, let $x_n \to x$ and $y_n \to y$, where $x_n, y_n \in X$. Then, for scalars λ and μ, the sequence $(\lambda x_n + \mu y_n)_1^\infty$ is a Cauchy sequence. Now, if $\tilde{X}$ is to be given a normed-space structure such that the norm on $\tilde{X}$ induces the metric on $\tilde{X}$, then

$$\lim_{n \to \infty} (\lambda x_n + \mu y_n) = \lambda \lim_{n \to \infty} x_n + \mu \lim_{n \to \infty} y_n = \lambda x + \mu y.$$

Thus $\lambda x + \mu y$ must be defined to be the limit of the Cauchy sequence $(\lambda x_n + \mu y_n)_1^\infty$. It is easily checked that with this definition of addition and scalar multiplication $\tilde{X}$ becomes a vector space. Furthermore, $\|x\| = d(x, 0) = \lim_{n \to \infty} \|x_n\|$ is a norm on this vector space, turning $\tilde{X}$ into a Banach space.

The remaining assertions are clear.   □

In fact, Theorem 7 is also an immediate consequence of Theorem 3.10. The space $\overline{X}$ in this result is called the *completion* of the normed space X. In view of Theorem 7, if X is a dense subspace of a Banach space Y then Y is often regarded as the completion of X. For example, the space $C[0, 1]$ of continuous functions on $[0, 1]$ is the completion of the space of piecewise linear functions on $[0, 1]$ with supremum norm

$$\|f\|_\infty = \sup\{|f(x)| : 0 \leq x \leq 1\} = \max\{|f(x)| : 0 \leq x \leq 1\}.$$

The same space $C[0, 1]$ is also the completion of the space of polynomials and of the space of infinitely differentiable functions. Similarly, for $1 \leq p < \infty$, $L_p(0, 1)$ is the completion of the space of polynomials with the norm

$$\|f\|_p = \left(\int_0^1 |f(x)|^p \, dx \right)^{1/p},$$

and also of the space of piecewise linear functions with the same norm.

When working in Banach spaces, one often considers series. Given a normed space X, a series

$$\sum_{n=1}^{\infty} x_n \quad (x_n \in X)$$

is said to the *convergent* to $x \in X$ if

$$\sum_{n=1}^{N} x_n \to x \quad \text{as } N \to \infty,$$

i.e. if

$$\lim_{N \to \infty} \left\| x - \sum_{n=1}^{N} x_n \right\| = 0.$$

We also say that x is the *sum* of the x_n and write

$$x = \sum_{n=1}^{\infty} x_n.$$

A series $\sum_{n=1}^{\infty} x_n$ is said to the *absolutely convergent* if $\sum_{n=1}^{\infty} \|x_n\| < \infty$. In a Banach space X every absolutely convergent series $\sum_{n=1}^{\infty} x_n$ is convergent. Indeed, let $y_N = \sum_{n=1}^{N} x_n$ be the Nth partial sum. We have to show that $(y_N)_1^{\infty}$ is a Cauchy sequence. Given $\epsilon > 0$, choose an n_0 such that $\sum_{n=n_0}^{\infty} \|x_n\| < \epsilon$. Then for $n_0 \leqslant N < M$ we have

$$\|y_N - y_M\| = \left\| \sum_{n=N+1}^{M} x_n \right\| \leqslant \sum_{n=N+1}^{M} \|x_n\| < \epsilon.$$

Hence $(y_N)_1^{\infty}$ is indeed a Cauchy sequence and so converges to some $x \in X$, so that $x = \sum_{n=1}^{\infty} x_n$.

In an incomplete space X the assertion above always fails for some absolutely convergent series $\sum_{n=1}^{\infty} x_n$: this is often useful in checking whether a space is complete or not.

Before showing this, let us mention a very useful trick when dealing with Cauchy sequences in metric spaces: we may always assume that we are dealing with a 'fast' Cauchy sequence. To be precise, given a Cauchy sequence $(x_n)_1^{\infty}$, *a priori* all we know is that there is a sequence $(n_k)_1^{\infty}$, say $n_k = 2^k$, such that $d(x_n, x_m) < 1/k$ if $n, m \geqslant n_k$. However, we may always assume that our sequence is much 'better' than this: $d(x_n, x_m) < 2^{-n}$ if $m \geqslant n$, or $d(x_n, x_m) < 2^{-2^n}$ if $m \geqslant n$, or whatever suits us. Indeed, let $f(n) > 0$ be an arbitary function. Set $\epsilon_1 = f(1)$ and choose an n_1 such that $d(x_n, x_m) < \epsilon_1$ whenever $n, m \geqslant n_1$. Then set $\epsilon_2 = f(2)$ and choose an $n_2 > n_1$ such that $d(x_n, x_m) < \epsilon_2$ whenever

$n, m \geq n_2$. Continuing in this way, we find a sequence $n_1 < n_2 < \cdots$ such that $d(x_{n_k}, x_{n_l}) < f(k)$ if $l \geq k$. Setting $y_k = x_{n_k}$, we find that the subsequence $(y_k)_1^\infty$ is such that $d(y_k, y_l) < f(k)$ whenever $k < l$. Now if we can show that $(y_k)_1^\infty$ converges to some limit y, then $(x_n)_1^\infty$ tends to the same limit y. Indeed, if $d(x_n, x_m) < \epsilon$ for $n, m \geq n_0$ then for $n \geq n_0$ we have

$$d(x_n, y) = d(x_n, \lim_{k \to \infty} y_k) \leq \limsup_{k \to \infty} d(x_n, y_k) \leq \epsilon,$$

since $y_k = x_{n_k}$ and $n_k \geq n_0$ if k is sufficiently large. Thus, in proving that a Cauchy sequence $(x_n)_1^\infty$ in a metric space is convergent, we may indeed assume that $d(x_n, x_m) < 2^{-n}$, say, if $m \geq n$.

Theorem 8. A normed space is complete if and only if every absolutely convergent series in it is convergent.

Proof. We have already seen that in a Banach space every absolutely convergent series is convergent. Suppose then that in our space X every absolutely convergent series is convergent. Our aim is to show that every Cauchy sequence is convergent.

Let $(x_n)_1^\infty$ be a Cauchy sequence in X. As we have just seen, we may assume that $d(x_n, x_m) = \|x_n - x_m\| < 2^{-n}$ for $n \leq m$. Set $x_0 = 0$ and $y_k = x_k - x_{k-1}$ $(k \geq 1)$. Then $x_n = \sum_{k=1}^n y_k$ and $\|y_k\| < 2^{-k+1}$. Therefore the series $\sum_{k=1}^\infty y_k$ is absolutely convergent and has partial sums $x_1, x_2, \ldots$. By assumption, $\sum_{k=1}^\infty y_k$ is convergent, say to a vector x, and so $x_n \to x$. $\square$

Absolutely convergent series in Banach spaces have many properties analogous to those of absolutely convergent series in $\mathbb{R}$. For example, if $\sum_{n=1}^\infty x_n$ is absolutely convergent to x and $n_1, n_2, \ldots$ is a permutation of $1, 2, \ldots$ then $\sum_{i=1}^\infty x_{n_i}$ is also absolutely convergent to x. However, as we shall see in a moment, unlike in $\mathbb{R}$, this property does *not* characterize absolutely convergent series in a Banach space.

The use of series often enables one to identify a Banach space with a sequence space endowed with a particular norm. This identification is made with the aid of a basis, provided that the space does have a basis. Given a Banach space X, a sequence $(e_i)_1^\infty$ is said to be a (*Schauder*) *basis* of X if every $x \in X$ can be represented in the form $x = \sum_{i=1}^\infty \lambda_i e_i$ (λ_i scalar) and this representation is unique.

For $1 \leq p < \infty$ the space l_p has a basis; the so-called *standard* or *canonical basis* $(e_i)_1^\infty$, where $e_i = (0, \ldots, 0, 1, 0, \ldots) = (\delta_{1i}, \delta_{2i}, \ldots)$: the *i*th term is 1 and the others are 0. It is easily seen that if $x = (x_1, x_2, \ldots) \in l_p$ then $x = \sum_{i=1}^{\infty} x_i e_i$ and this representation is unique (see Exercise 11). In particular, $\sum_{i=1}^{\infty} x_i e_i$ is convergent (in l_p) if and only if $\sum_{i=1}^{\infty} |x_i|^p < \infty$. This implies that in l_2 every rearrangement of $\sum_{n=1}^{\infty} e_n/n$ is convergent (to the same sum), but the series is not absolutely convergent.

To conclude this chapter, let us say a few words about defining new spaces from old. We have already seen the simplest way: every (algebraic) subspace Y of a normed space X is a normed space, with the restriction of the norm of X to Y as the norm. If X is complete then Y is complete if and only if it is closed. As the intersection of a family of subspaces is again a subspace, for every set $S \subset X$ there is a unique smallest subspace containing S, called the *linear span of S* and denoted by $\text{lin} S$: it is the intersection of all subspaces containing S and also the set of all (finite) linear combinations of elements of S.

$$\text{lin} S = \bigcap \{W : W \text{ is a subspace of } X \text{ and } S \subset W\}$$

$$= \left\{ \sum_{i=1}^{n} \lambda_i s_i : s_1, \ldots, s_n \in S; \, n = 1, 2, \ldots \right\}.$$

Similarly, the *closed linear span* of S, denoted $\overline{\text{lin}} S$, is the unique smallest closed subspace containing S; it is the intersection of all closed subspaces containing S, and is also the closure of $\text{lin} S$, the linear span of S.

Let us turn now to quotient spaces. Given a vector space X and a subspace Z, define an equivalence relation $\sim$ on X by setting $x \sim y$ if $x - y \in Z$. For $x \in X$, let $[x]$ be the equivalence class of x: putting it another way, $[x] = x + Z$. Then $X/\sim = \{[x] : x \in X\}$ is a vector space, with vector-space operations induced by those on X: $\lambda[x] + \mu[y] = [\lambda x + \mu y]$. Note that $[x] = 0$ iff $x \in Z$. Now if X is a normed space and Z is a *closed* subspace of X then we can define a norm $\|\cdot\|_0$ on X/Z by setting

$$\|[x]\|_0 = \inf\{\|y\| : y \sim x\} = \inf\{\|x + z\| : z \in Z\}.$$

It is easily checked that $\|\cdot\|_0$ is indeed a norm on X/Z: the homogeneity and the triangle inequality are obvious and $\|[0]\|_0 = 0$. All that remains is to show that if $\|[x]\| = 0$ then $[x] = 0$. To see this, note that if $\|[x]\| = 0$ then $\|x - z_n\| \to 0$ for some sequence $(z_n) \subset Z$. Hence $z_n \to x$ and so, as Z is closed, $x \in Z$.

We call X/Z, endowed with $\|\cdot\|_0$, the *quotient normed space* of X by Z, and call $\|\cdot\|_0$ the *quotient norm*. Throughout this book, a quotient space of a normed spaces will always be endowed with the quotient norm.

Given normed spaces X and Y, and a bounded linear operator $T: X \to Y$, the kernel $Z = \mathrm{Ker}\, T = T^{-1}(0)$ of T is a closed subspace of X, and T induces a linear operator $T_0: X/Z \to Y$. Analogously to many standard results in algebra, we have the following theorem.

Theorem 9. Let X and Y be normed spaces, $T \in \mathscr{B}(X, Y)$ and $Z = \mathrm{Ker}\, T$. Let $T_0: X/Z \to Y$ be the linear map induced by T. Then T_0 is a bounded linear operator from the quotient space X/Z to Y, and its norm is precisely $\|T\|$.

Proof.

$$\|T_0[x]\| \leqslant \inf\{\|T\|\|y\| : y \sim x\} = \|T\|\|[x]\|,$$

showing that $\|T_0\| \leqslant \|T\|$.

Conversely, let $\epsilon > 0$ and choose an $x \in X$ such that $\|x\| = 1$ and $\|Tx\| > \|T\| - \epsilon$. Then $\|[x]\| \leqslant 1$ and $\|T_0[x]\| = \|Tx\| > \|T\| - \epsilon$. Hence $\|T_0\| > \|T\| - \epsilon$. $\qquad\square$

In fact, the quotient norm $\|\cdot\|_0$ on X/Z is the minimal norm on X/Z such that if $T \in \mathscr{B}(X, Y)$ and $\mathrm{Ker}\, T \supset Z$, then the operator $T_0: X/Z \to Y$ induced by T has norm at most $\|T\|$ (see Exercise 13).

Suppose that X and Y are closed subspaces of a normed space Z, with $X \cap Y = \{0\}$ and $X + Y = Z$. If the projections $p_X: Z \to X$, and $p_Y: Z \to Y$, given by $p_X(x, y) = x$ and $p_Y(x, y) = y$, are bounded (i.e. continuous) then we call Z a *direct sum* of the subspaces X and Y. It is easily seen that Z is a direct sum of its subspaces X and Y iff the topology on Z (identified with $X \oplus Y = \{(x, y): x \in X, y \in Y\}$) is precisely the product of the topologies on X and Y. Note that, if Z is a direct sum of X and Y, Z' is a direct sum of X' and Y', and X is isomorphic to X' and Y is isomorphic to Y', then Z is isomorphic to Z'. If Z is a direct sum of X and Y then the projection $p_X: Z \to X$ induces an isomorphism between Z/Y and X.

Conversely, given normed spaces X and Y, there are various natural ways of turning $X \oplus Y$, the algebraic direct sum of the underlying vector spaces, into a normed space. For example, for $1 \leqslant p \leqslant \infty$, we may take the norm $\|(x, y)\|_p = \|(\|x\|, \|y\|)\|_p$. Thus for $1 \leqslant p < \infty$ we take $\|(x, y)\|_p = (\|x\|^p + \|y\|^p)^{1/p}$, and for $p = \infty$ we define $\|(x, y)\|_\infty = \max\{\|x\|, \|y\|\}$. It is easily seen that all these norms are equivalent; indeed, each induces the product topology on $X \oplus Y$. The normed space $(X \oplus Y, \| \cdot \|_p)$ is usually denoted by $X \oplus_p Y$. Considering X and Y as subspaces of $X \oplus_p Y$, we see that $X \oplus_p Y$ is a direct sum of X and Y.

Finally, given a family $\{ \| \cdot \|_\gamma : \gamma \in \Gamma \}$ of norms on a vector space V, if

$$\|x\| = \sup_{\gamma \in \Gamma} \|x\|_\gamma < \infty$$

for every $x \in V$, then $\| \cdot \|$ is a norm on V. Note that the analogous assertion about the infimum of norms does not hold in general (see Exercise 15).

Having got a good many of the basic definitions under our belts, we are ready to examine the concepts in some detail. In the next chapter we shall study continuous linear functionals.

Exercises

1. Show that in a normed space, the closure of the open ball $D_r(x_0)$ $(r > 0)$ is the closed ball $B_r(x_0)$ and the boundary $\partial B_r(x_0)$ of $B_r(x_0)$ is the sphere $S_r(x_0)$. Do these statements hold in a general metric space as well?

2. Let $B_1 \supset B_2 \supset \cdots$ be closed balls in a normed space X, where

$$B_n = B(x_n, r_n) = \{x \in X : \|x_n - x\| \leq r_n\} \qquad (r_n > r > 0).$$

Does

$$\bigcap_{n=1}^{\infty} B_n \neq \varnothing$$

hold? Is there a ball $B(x,r)$ $r > 0$, contained in $\bigcap_{n=1}^{\infty} B_n$?

3. Prove or disprove each of the following four statements. In a complete $\genfrac{}{}{0pt}{}{\text{metric}}{\text{normed}}$ space every nested sequence of closed $\genfrac{}{}{0pt}{}{\text{balls}}{\text{bounded sets}}$ has a non-empty intersection.

4. Let $X = (V, \| \cdot \|)$ be a normed space and W a subspace of V. Suppose $| \cdot |$ is a norm on W which is equivalent to the restriction of $\| \cdot \|$ to W. Show that there is a norm $\| \cdot \|_1$ on V that is equivalent to $\| \cdot \|$ and whose restriction to W is precisely $| \cdot |$.

5. Let $\| \cdot \|$ and $| \cdot |$ be two norms on a vector space V and let W be a subspace of V that is $\| \cdot \|$-dense in V. Suppose that the restrictions of $\| \cdot \|$ and $| \cdot |$ to W are equivalent. Are $\| \cdot \|$ and $| \cdot |$ necessarily equivalent?

6. For $1 \leq p \leq \infty$, let $\| \cdot \|_p$ be the l_p-norm on $\mathbb{R}^n$. Show that if $1 \leq p < r \leq \infty$ then $\|x\|_p \geq \|x\|_r$. For which points x do we have equality?

Prove that for every $\epsilon > 0$ there is an N such that if $N < p < \infty$ then

$$\|x\|_\infty \leqslant \|x\|_p \leqslant (1+\epsilon)\|x\|_\infty.$$

7. Show that the space defined in Examples 1 (xix) is indeed a normed space.

8. Let $1 \leqslant p, q, r \leqslant \infty$ be such that $p^{-1} + q^{-1} + r^{-1} = 1$, with $1/\infty$ defined to be 0. Show that for $x, y, z \in \mathbb{C}^n$ we have

$$\left| \sum_{i=1}^{n} x_i y_i z_i \right| \leqslant \|x\|_p \|y\|_q \|z\|_r.$$

State and prove the analogous inequality for s vectors from $\mathbb{C}^n$.

9. Show that l_p is a Banach space for every p, $(1 \leqslant p \leqslant \infty)$ and that c_0, the set of all sequences tending to 0, is a closed subspace of l_∞. Show also that l_p $(1 \leqslant p < \infty)$ and c_0 are separable Banach spaces (i.e. each contains a countable dense set) while l_∞ is not separable.

10. Let p, q and r be positive reals satisfying $p^{-1} + q^{-1} = r^{-1}$. Show that for $f \in L_p(0,1)$ and $g \in L_q(0,1)$ the function fg belongs to $L_r(0,1)$ and

$$\|fg\|_r \leqslant \|f\|_p \|g\|_q.$$

11. Let $e_i = (0,\ldots,0,1,0,\ldots) = (\delta_{1i}, \delta_{2i}, \ldots) \in l_p$, where $1 \leqslant p < \infty$. Show that $(e_1, e_2, \ldots)$ is a basis of l_p, called the *standard basis*, i.e. every $x \in l_p$ has a unique representation in the form

$$x = \sum_{i=1}^{\infty} \lambda_i e_i.$$

12. For $x = (x_i)_1^\infty \in l^p$, the *support of* x is $\operatorname{supp} x = \{i : x_i \neq 0\}$. Let $1 \leqslant p < \infty$ and let $f_1, f_2, \ldots \in l_p$ be non-zero vectors with disjoint supports. Show that $X = \overline{\operatorname{lin}}(f_i)_1^\infty$ is isometric to l_p: in fact, the map $X \to l_p$ given by $f_i \mapsto \|f_i\|_p e_i$ defines a linear isometry.

13. Let Z be a closed subspace of a normed space X. Show that the quotient norm on the vector space X/Z is the minimal norm such that if Y is a normed space, $T \in \mathcal{B}(X, Y)$ and $Z \subset \operatorname{Ker} T$ then the norm of the induced operator $T_0 : X/Z \to Y$ is at most $\|T\|$.

14. A *seminorm* on a vector space V is a function $p : V \to \mathbb{R}^+$ such that $p(\lambda x) = |\lambda| p(x)$ and $p(x+y) \leqslant p(x) + p(y)$ for all vectors $x, y \in V$ and scalar λ. [Thus a seminorm p is a norm if $p(x) = 0$ implies $x = 0$.]

Let $\{p_\gamma : \gamma \in \Gamma\}$ be a family of seminorms on a vector space V such that

$$0 < p(x) = \sup\{p_\gamma(x) : \gamma \in \Gamma\} < \infty$$

for every $x \in V$ $(x \neq 0)$. Show that $p(\cdot)$ is a norm on V.

15. Let $\|x\|_1$ and $\|x\|_2$ be norms on a vector space V. Is $\|x\| = \min\{\|x\|_1, \|x\|_2\}$ necessarily a norm?

16. Consider the vector space c_0 of complex sequences $x = (x_n)_1^\infty$ tending to 0. For a sequence $(x_n)_1^\infty \in c_0$, let $(x_n^*)_1^\infty$ be the decreasing rearrangement of $(|x_n|)_1^\infty$. Formally, $x_n^* = x$ if

$$|\{k : |x_k| > x\}| < n \leq |\{k : |x_k| \geq x\}|.$$

Let $b_1 \geq b_2 \geq \cdots > 0$ be such that $\sum_{i=1}^\infty b_i = \infty$. Define, for $x = (x_n)_1^\infty \in c_0$,

$$\|x\|' = \sup\left\{\sum_{n=1}^m x_n^* \Big/ \sum_{n=1}^m b_n : m = 1, 2, \ldots\right\}.$$

Let

$$d_0 = \{x \in c_0 : \|x\|' < \infty\}.$$

Show that $\|\cdot\|'$ is a norm on d_0. Is this space complete?

17. For $x \in l_1$ set

$$\|x\|' = 2\left|\sum_{n=1}^\infty x_n\right| + \sum_{n=2}^\infty \left(1+\frac{1}{n}\right)|x_n|.$$

Show that $\|\cdot\|'$ is a norm on l_1 and that l_1 is complete in this norm. Is $\|x\|'$ equivalent to the l_1-norm $\|x\|_1 = \sum_{n=1}^\infty |x_n|$?

18. Show that on every infinite-dimensional normed space X there is a discontinuous (unbounded) linear functional. [By Zorn's lemma X has a *Hamel basis*, i.e. a set $\{x_\gamma : \gamma \in \Gamma\} \subset X$ such that every $x \in X$ has a unique representation in the form $x = \sum_{i=1}^n \lambda_i x_{\gamma_i}$ $(\gamma_i \in \Gamma; 1 \leq i \leq n; n \in \mathbb{N})$.]

19. Prove that if two norms on the same vector space are not equivalent then at least one of them is discontinuous on the unit sphere in the other norm. Can each norm be discontinuous when restricted to the unit sphere of the other?

20. Give two norms on a vector space such that one is complete and the other is incomplete.

21. Find two inequivalent norms $\|\cdot\|_1$ and $\|\cdot\|_2$ on a vector space V such that $(V, \|\cdot\|_1)$ and $(V, \|\cdot\|_2)$ are isometric normed spaces.

22. Let Y be a closed subspace of a Banach space X and let x be an element of X. Is the distance of x from Y attained? (Is there a point $y_0 \in Y$ such that $\|x-y_0\| = \inf\{\|x-y\| : y \in Y\}$?)

23. Let $f_1, f_2, \ldots, f_n$ be linear functionals on a vector space V. Show that there is a norm $\|\cdot\|$ on V such that each f_i is continuous on $(V, \|\cdot\|)$. Can this be done for infinitely many linear functionals? And what about infinitely many linearly independent linear functionals?

24[+]. Let X be a Banach space. Suppose $(x_n)_1^\infty \subset X$ is a sequence such that every $x \in X$ has a unique representation in the form $x = \sum_{n=1}^\infty \lambda_n x_n$. Prove that the set $\{x_n\}_1^\infty$ consists of isolated points.

25. Check the assertion in Examples 6 (x).

26. Let Y be a closed subspace of a normed space X. Show that if X/Y and Y are separable then so is X.

27. Let Y be a closed subspace of a normed space X. Show that if any two of the three spaces X, Y and X/Y are complete then so is the third.

28. Let Y be a subspace of a normed space X. Show that Y is a closed subspace if and only if its unit ball, $B(Y)$, is closed in X. Show also that if Y is complete then Y is closed.

29. Prove the converse of one of the assertions of Theorem 4: if $\mathcal{B}(X, Y)$ is complete then so is Y.

30[+]. Let V be a vector space with basis $(x_n)_1^\infty$. [Thus every $x \in V$ $(x \neq 0)$ has a unique expression in the form $x = \sum_{i=1}^k \lambda_i x_{n_i}$ $(n_1 < n_2 < \cdots < n_k$ and $\lambda_i \neq 0).$] Show that there is no complete norm on V.

31[+]. Let C be a closed convex set in a normed space X such that $C + B(X) \supset B_{1+\epsilon}(X)$ for some $\epsilon > 0$. Does it follow that $\operatorname{Int} C \neq \emptyset$, i.e. that C contains a ball of positive radius?

Notes

Abstract normed spaces were first defined and investigated by Stefan Banach in 1920 in his Ph.D. thesis at the University of Léopol (i.e. the Polish town of Lwów, now in the Soviet Union). Much of this thesis was published as an article: *Sur les opérations dans les ensembles abstraits et leur applications aux équations intégrales*, Fundamenta Mathematica, 3 (1922), 133–81. A few years later Banach wrote the first book wholly devoted to normed spaces and linear operators: *Théorie des Opérations Linéaires*, Warsaw, 1932, vii + 254 pp. Banach gave an elegant account of the work of many mathematicians involved in the creation of functional analysis, including Frédéric (Frigyes) Riesz, whom he quoted most frequently, just ahead of himself, Maurice

Fréchet, Alfréd Haar, Henri Lebesgue, Stanislaw Mazur, Juliusz Schauder and Hugo Steinhaus. It is perhaps amusing to note that, when writing about Banach spaces, Banach used the term 'espace du type (*B*)'.

This beautiful book of Banach has had a tremendous influence on functional analysis; it is well worth reading even today, expecially in its English translation: *Theory of Linear Operators* (translated by F. Jellett), North-Holland, Amsterdam, 1987, ix + 237 pp. This edition is particularly valuable because the second part, by A. Pelczyński and Cz. Bessaga (*Some aspects of the present theory of Banach spaces*, pp. 161–237), brings the subject up to date, with many recent results and references.

Another classic on functional analysis is F. Riesz and B. Sz.-Nagy, *Functional Analysis* (translated from the 2nd French edition by L. F. Boron), Blackie and Son Ltd., London and Glasgow, 1956, xii + 468 pp. This volume concentrates on the function-theoretic and measure-theoretic aspects of functional analysis, so it does not have too much in common with our treatment of linear analysis.

There are a good many monographs on normed spaces and linear operators, including the massive treatise by N. Dunford and J. Schwartz, *Linear Operators*, in three parts; Interscience, New York; Part I: General Theory, 1958, xiv + 858 pp.; Part II: Spectral Theory, Self Adjoint Operators in Hilbert Space, 1963, ix + 859–1923 pp. + 7 pp. Errata; Part III: Spectral Operators, 1971, xix + 1925–2592 pp., L. V. Kantorovich and G. P. Akilov, *Functional Analysis in Normed Spaces*, Pergamon Press, International Series of Monographs on Pure and Applied Mathematics, vol. 45, Oxford, 1964, xiii + 771 pp., A. N. Kolmogorov and S. V. Fomin, *Introductory Real Analysis* (translated and edited by R. A. Silverman), Prentice-Hall, Inc., Englewood Cliffs, N. J., 1970, xii + 403 pp., Mahlon M. Day, *Normed Linear Spaces*, Third Edition, Ergebnisse der Mathematik und Ihrer Grenzgebiete, vol. 21, Springer-Verlag, Berlin, 1973, viii + 211 pp. and J. B. Conway, *A Course in Functional Analysis*, Graduate Texts in Mathematics, vol. 96, Springer-Verlag, New York, 1985, xiv + 404 pp.

3. LINEAR FUNCTIONALS AND THE HAHN–BANACH THEOREM

Given a normed space $X = (V, \|\cdot\|)$, let us write X' for the *algebraic dual* of X, i.e. for the vector space V' of linear functionals on V. Thus X^*, the space of bounded linear functionals on X, is a subspace of the vector space X'.

We know from the standard theory of vector spaces that every independent set of vectors is contained in a (Hamel) basis (see Exercise 18 of Chapter 2). In particular, for every non-zero vector $u \in V$ there is a linear functional $f \in V'$ with $f(u) \neq 0$. Equivalently, V' is a large enough to distinguish the elements of V: for all $x, y \in V$ ($x \neq y$) there is a functional $f \in V'$ such that $f(x) \neq f(y)$. Even more, the dual V'' of V' is large enough to accommodate V: there is a natural embedding of V into V'' which is an isomorphism if V is finite-dimensional.

But what happens if we restrict our attention to *bounded* linear functionals on a normed space X? Are there sufficiently many bounded linear functionals to distinguish the elements of X? In other words, given an element $x \in X$ ($x \neq 0$) is there a functional $f \in X^*$ such that $f(x) \neq 0$? As X is a normed space, one would like to use X^* to obtain some information about the norm on X. So can we estimate $\|x\|$ by $f(x)$ for some $f \in X^*$ with $\|f\| = 1$? To be more precise, we know that for every $x \in X$,

$$\|x\| \geq \sup\{|f(x)| : f \in B(X^*)\}.$$

But is the right-hand-side comparable to $\|x\|$?

As a matter of fact, so far we do not even know that for every non-zero normed space there is at least one non-zero linear functional, i.e. we have not even ruled out the utter indignity that $X^* = \{0\}$ while $X = (V, \|\cdot\|)$ is large, say V is infinite-dimensional.

The main aim of this chapter is to show that, as far as the questions above are concerned, Candide and Pangloss were right, *tout est pour le mieux dans le meilleur des mondes possibles*; indeed, everything is for the best in the world of bounded linear functionals. Before we present the result implying this, namely the Hahn–Banach theorem, we shall point out some elementary facts concerning linear functionals.

Let us show that $f \in X'$ is bounded if and only if $f(B)$ is not the entire ground field. Let $B = B(X) = B(0,1)$ be the closed unit ball of X. If $|\lambda| \leqslant 1$ then $\lambda B = B(0,|\lambda|) \subset B$. Hence for $f \in X'$ we have $\sup|f(B)| = \sup\{|f(x)| : \|x\| \leqslant 1\} = \|f\|$, where $\|f\| = \infty$ if f is unbounded, and $\lambda f(B) = f(\lambda B) \subset f(B)$. Consequently $f(B)$ is either $\{\lambda : |\lambda| < \|f\|\}$ or $\{\lambda : |\lambda| \leqslant \|f\|\}$. Similarly, if $D = D(X) = D(0,1)$ is the open unit ball of X then $f(D) = \{\lambda : |\lambda| < \|f\|\}$ for all $f \in X'$ $(f \neq 0)$. Hence $f \in X'$ is bounded iff $f(B)$ is not the entire ground field, as claimed.

It is often useful to think of a (non-zero) linear functional as a hyperplane in our vector space. An *affine hyperplane* or simply a *hyperplane* H in X is a set

$$H = \{x_0\} + Y = \{x_0 + y : y \in Y\},$$

where $x_0 \in X$ and $Y \subset X$ is a subspace of codimension 1, i.e. a subspace with $\dim X/Y = 1$. We say that H is a *translate* of Y. Given a non-zero functional $f \in X'$, let $K(f) = f^{-1}(0) = \{x \in X : f(x) = 0\}$ be the null space, i.e. the kernel, of f and let

$$I(f) = f^{-1}(1) = \{x \in X : f(x) = 1\}.$$

Let us recall the following simple facts from elementary linear algebra.

Theorem 1. Let X be a (real or complex) vector space.

(a) If $f \in X'$ and $f(x_0) \neq 0$, then $K(f)$ is a subspace of codimension 1. Moreover, if $f(x_0) \neq 0$ then every vector $x \in X$ has a unique representation in the form $x = y + \lambda x_0$, where $y \in K(f)$ and λ is a scalar.

    Furthermore, $I(f)$ is a hyperplane not containing 0.

(b) If $f, g \in X' - \{0\}$ then $f = \lambda g$ iff $K(f) = K(g)$.

(c) The map $f \mapsto I(f)$ gives a 1–1 correspondence between non-zero linear functionals and hyperplanes not containing 0. $\qquad\square$

Continuous (i.e. bounded) linear functionals are easily characterized in terms of $K(f)$ or $I(f)$. Note that $\|f\| \leqslant 1$ if and only if $|f(x)| < 1$ for all x with $\|x\| < 1$, i.e. if and only if $I(f)$ is disjoint from the open unit ball $D(0,1) = \{x \in X : \|x\| < 1\}$.

A subset A of a topological space T is *nowhere dense* if its closure has empty interior.

Theorem 2. Let X be a (real or complex) normed space.

(a) Let $f \in X'$, $(f \neq 0)$. If f is continuous (i.e. $f \in X^*$) then $K(f)$ and $I(f)$ are closed and nowhere dense in X. If f is discontinuous (i.e. unbounded) then $K(f)$ and $I(f)$ are dense in X.

(b) The map $f \mapsto I(f)$ gives a 1–1 correspondence between non-zero bounded linear functionals and closed hyperplanes not containing 0.

Proof. (a) Suppose that f is continuous. Then $K(f)$ and $I(f)$ are closed, since they are inverse images of closed sets. If $f(x_0) \neq 0$ then $f(x) \neq f(x + \epsilon x_0)$ for $\epsilon > 0$. Hence $K(f)$ and $I(f)$ have empty interiors, and so are nowhere dense.

Suppose now that $K(f)$ is not dense in X, say $B(x_0, r) \cap K(f) = \emptyset$ for some $x_0 \in X$ and $r > 0$. Then $f(B(x_0, r)) = f(x_0) + rf(B(X))$ does not contain 0, so $f(B(X))$ is not the entire ground field. Hence, as we have seen, f is bounded.

Since $I(f)$ is a translate of $K(f)$, it is dense iff $K(f)$ is dense.

(b) This is immediate from (a) and Theorem 1 (c). □

In fact, $B(x_0, r) \cap K(f) = \emptyset$ implies a bound on the norm of f, namely the bound $\|f\| \leq |f(x_0)|/r$. Indeed, otherwise $|f(x)| > |f(x_0)| \|x\|/r$ for some $x \in X$ and so

$$ y = x_0 - \frac{xf(x_0)}{f(x)} \in B(x_0, r) $$

and $f(y) = 0$, contradicting our assumption.

Now we turn to one of the cornerstones of elementary functional analysis, the Hahn–Banach theorem which guarantees that functionals can be extended from subspaces without increasing their norms. This means that all the questions posed at the beginning of the chapter have reassuring answers. Although the proof of the general form of the Hahn–Banach theorem uses Zorn's lemma, the essential part of the proof is completely elementary and very useful in itself.

Let $Y \subset X$ be vector spaces and let $f \in X'$ and $g \in Y'$. If $f(y) = g(y)$ for all $y \in Y$ (i.e. $f|Y$, the restriction of f to Y, is g) then f is an *extension* of g. We express this by writing $g \subset f$. A function $p : X \to \mathbb{R}^* = \mathbb{R} \cup \{\infty\}$ on a real vector space X is said to be a *convex functional* if it is positive homogeneous, i.e. $p(tx) = tp(x)$ for all $t \geq 0$

and $x \in X$, and is a convex function (as used in Chapter 1), i.e. if $x, y \in X$, and $0 \leqslant t \leqslant 1$ then $p(tx + (1-t)y) \leqslant tp(x) + (1-t)p(y)$. By the positive homogeneity of p, the second condition is equivalent to $p(x+y) \leqslant p(x) + p(y)$ for all $x, y \in X$, i.e. to the *subadditivity* of p. As customary for the operations on $\mathbb{R}^* = \mathbb{R} \cup \{\infty\}$, we use the convention that $\infty + s = \infty + \infty = \infty$ for all $s \in \mathbb{R}$; $0 . \infty = 0$; and $t . \infty = \infty$ for $t > 0$. Note that a norm is a convex functional, as is every linear functional. Furthermore, if $X = (V, \|\cdot\|)$ is a normed space then a linear functional $f \in X'$ is dominated by the convex functional $N\|x\|$ iff $f \in X^*$ and $\|f\| \leqslant N$. As usual, a function $\varphi : S \to \mathbb{R}$ is said to *dominate* a function $\psi : S \to \mathbb{R}$ if $\psi(s) \leqslant \varphi(s)$ for all $s \in S$.

Lemma 3. Let p be a convex functional on a real vector space X and let f_0 be a linear functional on a 1-codimensional subspace Y of X. Suppose that f_0 is dominated by p, i.e.

$$f_0(y) \leqslant p(y) \qquad \text{for all } y \in Y.$$

Then f_0 can be extended to a linear functional $f \in X'$ dominated by p:

$$f(x) \leqslant p(x) \qquad \text{for all } x \in X. \tag{1}$$

Proof. Fix $z \in X$ ($z \notin Y$) so that every $x \in X$ has a unique representation in the form $x = y + tz$, where $y \in Y$ and $t \in \mathbb{R}$. The functional f we are looking for is determined by its value on z, say $f(z) = c$. To prove (1), we have to show that for some choice of c we have

$$f(y + tz) \leqslant p(y + tz),$$

in other words

$$f_0(y) + tc \leqslant p(y + tz) \tag{2}$$

for all $y \in Y$ and $t \in \mathbb{R}$.

For $t > 0$ inequality (2) gives an upper bound on c, and for $t = -s < 0$ it gives a lower bound. Indeed, for $t > 0$, (2) becomes

$$c \leqslant \frac{p(y + tz) - f_0(y)}{t} = p\left(\frac{y}{t} + z\right) - f_0\left(\frac{y}{t}\right)$$

for all $y \in Y$. For $s > 0$ we have $f_0(y) - sc \leqslant p(y - sz)$ and so

$$c \geqslant \frac{-p(y - sz) + f_0(y)}{s} = -p\left(\frac{y}{s} - z\right) + f_0\left(\frac{y}{s}\right)$$

for all $y \in Y$. The former holds iff

$$c \leqslant p(y'+z) - f_0(y')$$

for all $y' \in Y$, and the latter holds iff

$$c \geqslant -p(y''-z) + f_0(y'')$$

for all $y'' \in Y$. Hence there is an appropriate c iff

$$f_0(y'') - p(y''-z) \leqslant p(y'+z) - f_0(y') \qquad (3)$$

for all $y', y'' \in Y$. But (3) does hold since

$$f_0(y') + f_0(y'') = f_0(y'+y'') \leqslant p(y'+y'') \leqslant p(y'+z) + p(y''-z),$$

completing the proof. $\qquad\qquad\qquad\qquad\qquad\qquad\qquad\qquad\qquad\qquad$ □

The following theorem is a slight strengthening of Lemma 3.

Theorem 4. Let Y be a subspace of a real vector space X such that X is the linear span of Y and a sequence $z_1, z_2, \ldots$. Suppose $f_0 \in Y'$ is dominated by a convex functional p on X. Then f_0 can be extended to a linear functional $f \in X'$ dominated by p.

If X is a real normed space and $f_0 \in Y^*$ then f_0 has an extension to a functional f on X such that $\|f\| = \|f_0\|$.

Proof. Set $X_n = \mathrm{lin}\{Y, z_1, \ldots, z_n\}$. By Lemma 3 we can define linear functionals $f_0 \subset f_1 \subset f_2 \subset \cdots$ such that $f_n \in X_n'$ and each f_n is dominated by p. Define $f: X \to \mathbb{R}$ by setting $f(x) = f_n(x)$ if $x \in X_n$. Then $f \in X'$ extends f_0 and it is dominated by p.

The second part is immediate from the first. Indeed, f_0 is dominated by the convex functional $p(x) = N\|x\|$ where $N = \|f_0\|$. Hence there is an $f \in X'$ extending f_0 and dominated by p. But then $f(x) \leqslant p(x) = N\|x\|$ for all $x \in X$, so that $\|f\| \leqslant N = \|f_0\|$, implying $\|f\| = \|f_0\|$. $\qquad\qquad$ □

The restriction on Y in Theorem 4 is, in fact, unnecessary. As we shall see, this is an easy consequence of Zorn's lemma, the standard weapon of an analyst which ensures the existence of maximal objects. For the sake of completeness, we shall state Zorn's lemma, but before doing so we have to define the terms needed in the statement.

A *partial order* or simply *order* on a set P is a binary relation $\leqslant$ such that (i) $a \leqslant a$ for every $a \in P$, (ii) if $a \leqslant b$ and $b \leqslant c$ for $a, b, c \in P$ then $a \leqslant c$, and (iii) if $a \leqslant b$ and $b \leqslant a$ for some $a, b \in P$ then $a = b$. Briefly, $\leqslant$ is a transitive and reflexive binary relation on P. We call the pair $(P, \leqslant)$ a *partially ordered set*; in keeping with our custom concerning normed spaces and topological spaces, $(P, \leqslant)$ is often abbreviated to P. A subset C of P is a *chain* or a *totally ordered set* if for all $a, b \in C$ we have $a \leqslant b$ or $b \leqslant a$. An element $m \in P$ is a *maximal element* of P if $m \leqslant a$ implies that $a = m$; furthermore, we say that b is an *upper bound* for a set $S \subset P$ if $s \leqslant b$ for all $s \in S$. It can be shown that the axiom of choice is equivalent to the following assertion.

Zorn's lemma. If every chain in a non-empty partially ordered set P has an upper bound, then P has at least one maximal element. $\square$

The fact that Theorem 4 holds for any subspace Y of X is the celebrated Hahn–Banach extension theorem.

Theorem 5. Let Y be a subspace of a real vector space X and let $f_0 \in Y'$. Let p be a convex functional on X. If f_0 is dominated by p on Y, i.e. $f_0(y) \leqslant p(y)$ for every $y \in Y$, then f_0 can be extended to a linear functional $f \in X'$ dominated by p.

If X is a real normed space and $f_0 \in Y^*$ then f_0 has a norm-preserving extension to the whole of X: there is a functional $f \in X^*$ such that $f_0 \subset f$ and $\|f\| = \|f_0\|$.

Proof. Consider the set $\mathcal{F} = \{f_\gamma : \gamma \in \Gamma\}$ of all extensions of f_0 dominated by p: for each γ there is a subspace Y_γ and a linear functional $f_\gamma \in Y_\gamma'$ such that $Y \subset Y_\gamma$, $f_0 \subset f_\gamma$ and f_γ is dominated by p. Clearly the relation '$\subset$' is a partial order on $\mathcal{F}$ (f_γ is 'less than or equal to' f_δ iff $f_\gamma \subset f_\delta$). If $\mathcal{F}_0 = \{f_\gamma : \gamma \in \Gamma_0\}$ is a non-empty chain (i.e. a totally ordered set) then it has an upper bound, namely $\bar{f} \in \bar{Y}'$, where $\bar{Y} = \bigcup_{\gamma \in \Gamma_0} Y_\gamma$ and $\bar{f}(y) = f_\gamma(y)$ if $y \in Y_\gamma$ ($\gamma \in \Gamma_0$). Therefore, by Zorn's lemma, there is a maximal extension. But by Lemma 3 every maximal extension is defined on the whole of X.

The second part follows as before. $\square$

With a little work one can show that norm-preserving extensions can be guaranteed in complex normed spaces as well. A complex normed space X can be considered as a real normed space; as such, we denote it by $X_{\mathbb{R}}$. We write $X_{\mathbb{R}}^*$ for the dual of $X_{\mathbb{R}}$. It is easily checked that the mapping $r : X^* \to X_{\mathbb{R}}^*$ defined by $r(f) = \mathrm{Re}\, f$ (i.e. $r(f)(x) = \mathrm{Re}\, f(x)$ for $x \in X$) is a one-to-one norm-preserving map onto $X_{\mathbb{R}}^*$. The inverse of r is the map $c : X_{\mathbb{R}}^* \to X^*$ defined by $c(g)(x) = g(x) - ig(ix)$. This enables us to deduce the complex form of the Hahn–Banach extension theorem.

Theorem 6. Let Y be a subspace of a complex normed space X and let $f_0 \in Y^*$. Then f_0 has a norm-preserving extension to the whole of X: there is a functional $f \in X^*$ such that $f_0 \subset f$ and $\|f\| = \|f_0\|$.

Proof. By Theorem 5, we can extend $r(f_0)$ to a functional g on $X_\mathbf{R}$ satisfying $\|g\| = \|r(f_0)\| = \|f_0\|$. The complex functional $f = c(g) \in X^*$ extends f_0 and satisfies $\|f\| = \|f_0\|$. $\qquad \square$

The Hahn–Banach theorem has many important consequences; we give some of them here.

Corollary 7. Let X be a normed space, and let $x_0 \in X$. Then there is a functional $f \in S(X^*)$ such that $f(x_0) = \|x_0\|$. In particular, $\|x_0\| \leq C$ iff $|g(x_0)| \leq C$ for all $g \in S(X^*)$.

Proof. We may assume that $x_0 \neq 0$. Let Y be the 1-dimensional subspace $\mathrm{lin}\{x_0\}$ and define $f_0 \in Y^*$ by $f_0(\lambda x_0) = \lambda \|x_0\|$. Then $\|f_0\| = 1$ and its extension f, guaranteed by the Hahn–Banach theorem, has the required properties. $\qquad \square$

Corollary 8. Let X be a normed space, and let $x_0 \in X$. If $f(x_0) = 0$ for all $f \in X^*$ then $x_0 = 0$. $\qquad \square$

The functional f whose existence is guaranteed by Corollary 7 is said to be a *support functional at* x_0. Note that if $x_0 \in S(X)$ and f is a support functional at x_0 then the hyperplane $I(f)$ is a *support plane* of the convex body $B(X)$ at x_0; in other words: $x_0 \in B(X) \cap I(f)$ and $I(f)$ contains no interior point of $B(X)$. The norm on X is said to be *smooth* if every $x_0 \in S(X)$ has a *unique* support functional.

Corollary 7 implies that the map $\mathcal{B}(X, Y) \to \mathcal{B}(Y^*, X^*)$, given by $T \mapsto T^*$, is an isometry, as remarked after Theorem 2.4, when we defined the adjoint.

Theorem 9. If X and Y are normed spaces and $T \in \mathcal{B}(X, Y)$ then $T^* \in \mathcal{B}(Y^*, X^*)$ and $\|T^*\| = \|T\|$.

Proof. As usual, we may and shall assume that X and Y are non-trivial spaces: $X \neq \{0\}$ and $Y \neq \{0\}$. We know that $\|T^*\| \leq \|T\|$. Given $\epsilon > 0$, there is an $x_0 \in S(X)$ such that $\|Tx_0\| \geq \|T\| - \epsilon$. Let $g \in S(Y^*)$ be a support functional at Tx_0: $g(Tx_0) = \|Tx_0\|$. Then

$$(T^*g)(x_0) = g(Tx_0) = \|Tx_0\| \geq \|T\| - \epsilon,$$

so that $\|T^*g\| \geq \|T\| - \epsilon$ and $\|T^*\| \geq \|T\| - \epsilon$. $\qquad \square$

Given a vector space V with dual V' and second dual $V'' = (V')'$, there is a natural embedding $V \to V''$ defined by $v \mapsto v''$, where v'' is defined by $v''(f) = f(v)$ for $f \in V$. Rather trivially, this embedding is an isomorphism if V is finite dimensional. If X is a normed space with dual X^*, second dual X^{**} and $x \in X$, then we write $\hat{x}$ for the restriction of x'' to X^*: $\hat{x}$ is the linear functional on X^* given by $\hat{x}(f) = f(x)$ for $f \in X^*$. In other words, with the bracket notation,

$$\langle \hat{x}, f \rangle = \langle f, x \rangle$$

for all $f \in X^*$. Since $|\hat{x}(f)| = |f(x)| \leq \|f\| \|x\|$, we have $\hat{x} \in X^{**}$ (not just $\hat{x} \in (X^*)'$), and moreover $\|\hat{x}\| \leq \|x\|$. The Hahn–Banach theorem implies that, in fact, we have equality here.

Theorem 10. The natural map $x \mapsto \hat{x}$ is a norm-preserving isomorphism (embedding) of a normed space X into its second dual X^{**}.

Proof. For $x \in X$ ($x \neq 0$), let f be a support functional at x: $\|f\| = 1$ and $f(x) = \|x\|$. Then $|\hat{x}(f)| = |f(x)| = \|x\|$ and $|\hat{x}(f)| \leq \|\hat{x}\| \|f\| = \|\hat{x}\|$, so that $\|x\| \leq \|\hat{x}\|$. □

In view of Theorem 10 it is natural to consider X as a subspace of X^{**}. If, under the natural identification, X is the whole of X^{**}, i.e. $X = X^{**}$, then X is said to be *reflexive*. We know that X^{**} is complete even when X is not, so a reflexive space is necessarily complete. However, a Banach space need not be reflexive. For example, l_p is reflexive for $1 < p < \infty$, and l_1 and l_∞ are not reflexive. Also, as we shall see, every finite-dimensional normed space is reflexive.

Writing $C(L)$ for the Banach space of bounded continuous functions on a topological space L with the uniform norm, it is easily seen that every Banach space X is a closed subspace of $C(L)$ for some metric space L. Indeed, put $L = B(X^*)$, and for $x \in X$ define $f_x = \hat{x}|L$. Then, by Theorem 10, the map $X \to C(L)$, given by $x \mapsto f_x$, is a linear isometry onto a closed subspace of $C(L)$. We shall see in chapter 8 that considerably more is true: instead of $C(L)$ we may take $C(K)$, the space of all continuous functions on a compact Hausdorff space K with the uniform norm.

To conclude this chapter, let us present a strengthening of the Hahn–Banach theorem. This time we wish to impose not only an upper bound but also a lower bound on our linear functional to be found, the upper bound being a convex functional, as before, and the lower bound a concave functional.

Given a real vector space X, a function $q : X \to \mathbb{R}_* = \mathbb{R}\cup\{-\infty\}$ is said to be a *concave functional* if $-q(x)$ is a convex functional, i.e. if q is positive homogeneous and $q(x+y) \geq q(x)+q(y)$, i.e., superadditive. Given a convex functional p and a concave functional q, our aim is to find a linear functional $f \in X'$ such that

$$q(x) \leq f(x) \leq p(x) \tag{4}$$

for all $x \in X$.

What condition does f have to satisfy on a subspace Y in order for f to be extendable? One's first guess is surely that

$$q(y) \leq f(y) \leq p(y)$$

for all $y \in Y$. While this condition is undoubtedly necessary, it need not be the whole story. Indeed, as $f(y) = f(x+y) - f(x)$ and $-f(x) \leq -q(x)$, we must have

$$f(y) \leq p(x+y) - q(x) \tag{5}$$

for all $y \in Y$ *and* $x \in X$. Inequality (5) is stronger than (4): putting $x = 0$ in (5) we find that $f(y) \leq p(y)$ for all $y \in Y$, and setting $x = -y$ we see that $f(y) \leq -q(-y)$, i.e. $q(-y) \leq f(-y)$ for all $y \in Y$. Of course, if $Y = X$ then conditions (4) and (5) are equivalent.

The following strengthening of the Hahn–Banach theorem shows that (5) is sufficient to guarantee the existence of an extension from Y to the whole of X.

Theorem 11. Let p be a convex functional and q a concave functional on a real vector space X, let Y be a subspace of X and let $f_0 \in Y'$ be such that

$$f_0(y) \leq p(x+y) - q(x)$$

for all $y \in Y$ and $x \in X$. Then f_0 has an extension $f \in X'$ such that

$$q(x) \leq f(x) \leq p(x)$$

for every $x \in X$.

Proof. The heart of the matter is the analogue of Lemma 3: once we have managed to extend f_0 to a slightly larger (i.e. one dimension larger) subspace, the rest follows as before.

Pick a vector $z \in X$ ($z \notin Y$) and set $Z = \text{lin}\{Y, z\}$. Let us show that f_0 has an extension f_1 to Z satisfying

$$f_1(u) \leq p(x+u) - q(x)$$

for all $u \in Z$ and $x \in X$. As in the proof of Lemma 3, we have to show that there is a suitable choice c for $f_1(z)$, i.e. that there is a $c \in \mathbb{R}$ such that

$$f_1(y+z) = f_0(y) + c \le p(x+y+z) - q(x)$$

and

$$f_1(y'-z) = f_0(y') - c \le p(x'+y'-z) - q(x')$$

for all $y, y' \in Y$ and $x, x' \in X$. Such a c exists if and only if

$$f_0(y') - p(x'+y'-z) + q(x') \le -f_0(y) + p(x+y+z) - q(x)$$

for all $y, y' \in Y$ and $x, x' \in X$. But this inequality does hold, since

$$f_0(y) + f_0(y') = f_0(y+y')$$
$$\le p(x+x'+y+y') - q(x+x')$$
$$\le p(x+y+z) + p(x'+y'-z) - q(x) - q(x').$$

Hence f_0 has a suitable extension f_1 to Z.

This assertion implies the analogue of Theorem 4, and an application of Zorn's lemma gives our theorem in its full generality.   □

Corollary 12. Let p be a convex functional and q a concave functional on a real vector space X such that q is dominated by p: for all $x \in X$ we have $q(x) \le p(x)$. Then there is a linear functional $f \in V'$ such that

$$q(x) \le f(x) \le p(x) \qquad \text{for all } x \in X.$$

Proof. Let $Y = (0) \subset X$. Then the trivial linear functional f_0 on Y satisfies the condition in Theorem 11, namely

$$f_0(0) \le p(x) - q(x)$$

for all $x \in X$. Hence f_0 has an extension to a linear functional $f \in X'$ such that $q(x) \le f(x) \le p(x)$ for all $x \in X$.   □

The following *separation theorem* is an easy consequence of Corollary 12.

Theorem 13. Let A and B be disjoint non-empty convex subsets of a real vector space X. Suppose that for some $\alpha \in A$ and every $x \in X$ there is an $\epsilon(x) > 0$ such that $\alpha + tx \in A$ for all t, $|t| \le \epsilon(x)$. Then A and B can be separated by a hyperplane, i.e. there is a non-zero linear functional $f \in X'$ and a real number c such that $f(x) \le c \le f(y)$ for all $x \in A$ and $y \in B$.

Proof. We may assume that $\alpha = 0$, i.e., for every $x \in X$ there is an $\epsilon = \epsilon(x) > 0$ such that $[-\epsilon x, \epsilon x] \subset A$. Define functions p and q on X by setting, for $x \in X$,

$$p(x) = \inf\{t \geqslant 0 : x \in tA\}; \qquad q(x) = \sup\{t \geqslant 0 : x \in tB\}.$$

It is easily checked that $p : X \to \mathbb{R}$ is a convex functional and $q : X \to \mathbb{R}_*$ is a concave functional. Furthermore, as $tA \cap tB = \varnothing$ for $t > 0$, we have $q(x) \leqslant p(x)$. Hence, by Corollary 12, there is a non-zero linear functional $f \in X'$ such that $q(x) \leqslant f(x) \leqslant p(x)$ for all $x \in X$. To complete the proof, note that if $x \in A$ and $y \in B$ then

$$f(x) \leqslant p(x) \leqslant 1 \leqslant q(y) \leqslant f(y).$$

Hence we may take $c = 1$. $\qquad\qquad\qquad\qquad\qquad\qquad\qquad\qquad$ $\square$

As the last result of this chapter, we shall show that the separation theorem gives a pleasant description of the closed convex hull of a set in a normed space. The *convex hull* $\mathrm{co}\, S$ of a set S in a vector space X is the intersection of all convex subsets of X containing S, so it is the unique smallest convex set containing S. Clearly,

$$\mathrm{co}\, S = \left\{ \sum_{i=1}^{n} t_i x_i : x_i \in S, t_i \geqslant 0 \; (i = 1, \ldots, n), \sum_{i=1}^{n} t_i = 1 \; (n = 1, 2, \ldots) \right\}.$$

If X is a *normed* space then the *closed convex hull* $\overline{\mathrm{co}}\, S$ of S is the intersection of all closed convex subsets of X containing S, so it is the unique smallest closed convex set containing S. As the closure of a convex set is convex, $\overline{\mathrm{co}}\, S$ is the closure of $\mathrm{co}\, S$.

The following immediate consequence of the separation theorem shows that $\overline{\mathrm{co}}\, S$ is the intersection of all closed half-spaces containing S. It is, of course, trivial that $\overline{\mathrm{co}}\, S$ is contained in this intersection.

Theorem 14. Let S be a non-empty subset of a real normed space X. Then $\overline{\mathrm{co}}\, S = \{x \in X : f(x) \leqslant \sup_{s \in S} f(s) \text{ for all } f \in X^*\}$.

Proof. Suppose that $x_0 \notin \overline{\mathrm{co}}\, S$. Then $B(x_0, r) \cap \overline{\mathrm{co}}\, S = \varnothing$ for some $r > 0$. Let $f \in X'$ ($f \neq 0$) separate the convex sets $B(x_0, r)$ and $\overline{\mathrm{co}}\, S$:

$$f(x) \geqslant c \geqslant f(y)$$

for all $x \in B(x_0, r)$ and $y \in \overline{\mathrm{co}}\, S$. Since the restriction of f to $B(x_0, r)$ is bounded above, $f \in X^*$ and since $f \neq 0$, $f(x_0) > c$. $\qquad\qquad$ $\square$

Throughout the book, we shall encounter many applications of the Hahn–Banach theorem and its variants. For example, the last result will be used in chapter 8.

Exercises

1. Let p and q be conjugate indices, with $1 \leqslant p < \infty$. Prove that $l_p^* = l_q$. Show also that $c_0^* = l_1$. [Note that this gives a quick proof of the fact that for $1 \leqslant p \leqslant \infty$ the space l_p is complete.]

2. Let c be the subspace of l_∞ consisting of all convergent sequences. What is the general form of a bounded linear functional on c?

3. Let p and q be conjugate indices, with $1 \leqslant p < \infty$. Prove that the dual of $L_p(0,1)$ is $L_q(0,1)$.

4. Check that for $1 < p < \infty$, the space l_p is reflexive. Check also that l_1, l_∞ and c_0 are not reflexive.

5. Let X and Y be normed spaces, and set $Z = X \oplus Y$, with norm $\|z\| = \|(x,y)\| = \|x\| + \|y\|$. What is Z^*?

6. Let $X_1, X_2, \ldots$ be normed spaces and let $X = \bigoplus_{n=1}^{\infty} X_n$ be the space of all sequences $x = (x_1, x_2, \ldots)$ which are eventually zero: $x_n \in X_n$ and $x_n = 0$ if n is sufficiently large, with pointwise operations and norm

$$\|x\| = \left(\sum_{n=1}^{\infty} \|x_n\|^2 \right)^{1/2}.$$

What is X^*?

7. Let X be a Banach space. Show that if $X^{***} = X^*$ then X is reflexive, i.e. $X^{**} = X$.

8. A linear functional $f \in l_\infty'$ is *positive* if $f(x) \geqslant 0$ when every $x = (x_n)_1^\infty \in l_\infty$ with $x_n \geqslant 0$ for all n. Show that every positive linear functional on l_∞ is bounded.

9. Show that $p_0(x) = p_0((x_n)_1^\infty) = \limsup_{n \to \infty} x_n$ is a convex functional on l_∞ and deduce the existence of a linear functional $f: l_\infty \to \mathbb{R}$ such that

$$\liminf_{n \to \infty} x_n \leqslant f(x) \leqslant \limsup_{n \to \infty} x_n$$

for every $x = (x_n)_1^\infty \in l_\infty$.

10. Let X be a complex normed space and let Y be a subspace of $X_\mathbb{R}$. (Thus if $y_1, y_2 \in Y$ and $r_1, r_2 \in \mathbb{R}$ then $r_1 y_1 + r_2 y_2 \in Y$.) Let $f_0 \in Y^*$ (i.e. let f_0 be a bounded *real* linear functional on Y) such that

that if $y_1, y_2, \lambda_1 y_1 + \lambda_2 y_2 \in Y$ for some $\lambda_1, \lambda_2 \in \mathbb{C}$ and $y_1, y_2 \in Y$ then $f_0(\lambda_1 y_1 + \lambda_2 y_2) = \lambda_1 f_0(y_1) + \lambda_2 f_0(y_2)$. Show that f_0 need not have an extension to a bounded complex linear functional on the whole of X, i.e. it need not have an extension to a functional $f \in X^*$.

11$^+$. Let X be a Banach space, Y a 1-codimensional subspace of X and X_1 and X_2 dense subspaces of X. Is $X_1 \cap X_2$ dense in X? Is $X_1 \cap Y$ dense in Y? What are the answers if X_1 and X_2 have codimension 1?

12. Let X be a normed space, Y a dense subspace of X and Z a closed finite-codimensional subspace of X. Is $Z \cap Y$ dense in Z?

13. Let Y be a subspace of a normed space X. Show that the closure of Y is

$$\bar{Y} = \bigcap \{\mathrm{Ker}\, f : f \in X^*, Y \subset \mathrm{Ker}\, f\}.$$

14. Let K be a closed convex set in a real normed space X. Show that every boundary point of K has a support functional: for every $x_0 \in \partial K$ there is an $f \in X^*$ such that $f \neq 0$ and $\sup_{x \in K} f(x) = f(x_0)$.

15. Let A be a set of points in a real normed space X, and let $f_0 : A \to \mathbb{R}$. Show that there is a functional $f \in B(X^*)$ such that $f(a) = f_0(a)$ for all $a \in A$ if and only if

$$\left| \sum_{a \in F} \lambda(a) f_0(a) \right| \leq \left\| \sum_{a \in F} \lambda(a) a \right\|$$

for every finite subset F of A and for every function $\lambda : F \to \mathbb{R}$.

16. Let Y be a subspace of a normed space X and let $x \in X$. Show that

$$d(x, Y) = \inf\{\|x - y\| : y \in Y\} \geq 1$$

if and only if there is a linear functional $f \in B(X^*)$ such that $Y \subset \mathrm{Ker}\, f$ and $f(x) = 1$.

The results in the final exercises, all due to Banach, enable us to define finitely additive 'integrals' of large classes of functions and to attach a 'limit' to every bounded sequence.

17. Let $\mathbb{T} = \mathbb{R}/\mathbb{Z}$ be the circle group, i.e. the additive group of reals modulo the integers. Let $X = \mathscr{F}_b(\mathbb{T})$ be the vector space of bounded real-valued functions on $\mathbb{T}$. For

$$f = f(t) \in X \qquad \text{and} \qquad \alpha_1, \ldots, \alpha_n \in \mathbb{R}$$

set

$$r(f; \alpha_1, \ldots, \alpha_n) = \sup_{-\infty < t < \infty} \frac{1}{n} \sum_{i=1}^{n} f(t + \alpha_i)$$

and define

$$p(f) = \inf\{r(f; \alpha_1, \ldots, \alpha_n) : n = 1, 2, \ldots; \alpha_1, \ldots, \alpha_n \in \mathbb{R}\}.$$

Show that p is a convex functional and deduce that there is a generalized 'integral' $I(f)$ on $\mathcal{F}_b(\mathbb{T})$ such that for $f, g \in \mathcal{F}_b(\mathbb{T})$ and $\lambda, \mu, t_0 \in \mathbb{R}$ we have

(i) $I(\lambda f + \mu g) = \lambda I(f) + \mu I(g)$;   (ii) $I(f) \geqslant 0$ if $f \geqslant 0$;

(iii) $I(f(t + t_0)) = I(f(t))$;   (iv) $I(f(-t)) = I(f(t))$;

(v) $I(1) = 1$ (i.e. the integral of the identically 1 function is 1).

18. Let X be the vector space of all real-valued functions $f(t)$ on $\mathbb{R}$ such that $\limsup_{t \to \infty} |f(t)| < \infty$. For

$$f = f(t) \in X \qquad \text{and} \qquad \alpha_1, \ldots, \alpha_n \in \mathbb{R}$$

set

$$r(f; \alpha_1, \ldots, \alpha_n) = \limsup_{t \to \infty} \frac{1}{n} \sum_{i=1}^{n} f(t + \alpha_i)$$

and define

$$p(f) = \inf\{r(f; \alpha_1, \ldots, \alpha_n) : n = 1, 2, \ldots; \alpha_1, \ldots, \alpha_n \in \mathbb{R}\}.$$

Show that p is a convex functional, and deduce that there is a generalized 'limit' $\mathrm{LIM}_{t \to \infty} f(t)$ on X such that

(i) $\mathrm{LIM}_{t \to \infty}\{\lambda f(t) + \mu g(t)\} = \lambda \mathop{\mathrm{LIM}}_{t \to \infty} f(t) + \mu \mathop{\mathrm{LIM}}_{t \to \infty} g(t)$;

(ii) $\mathop{\mathrm{LIM}}_{t \to \infty} f(t) \geqslant 0$ if $\liminf_{t \to \infty} f(t) \geqslant 0$;

(iii) $\mathop{\mathrm{LIM}}_{t \to \infty} f(t + t_0) = \mathop{\mathrm{LIM}}_{t \to \infty} f(t)$;

(iv) $\mathop{\mathrm{LIM}}_{t \to \infty} 1 = 1$.

19. For $x = (x_n)_1^\infty \in l_\infty$ and $n_1 < \cdots < n_k$ put

$$\pi(x; n_1, \ldots, n_k) = \limsup_{n \to \infty} \frac{1}{k} \sum_{i=1}^{k} x_{n+n_i}$$

and define

$$p(x) = \inf\{\pi(x; n_1, \ldots, n_k) : n_1 < \cdots < n_k; k = 1, 2, \ldots\}.$$

Show that p is a convex functional, and deduce the existence of a linear functional $L: l_\infty \to \mathbb{R}$ such that

$$L(x) \leqslant p(x)$$

for every $x \in l_\infty$. The value $L(x)$ is said to be a *Banach limit* or a *generalized limit* of the sequence (x_n) and is usually denoted by $\mathrm{LIM}_{n\to\infty} x_n$ or $\mathrm{Lim}_{n\to\infty} x_n$.

Show that

(i) $\displaystyle \mathop{\mathrm{LIM}}_{n\to\infty} (\lambda x_n + \mu y_n) = \lambda \mathop{\mathrm{LIM}}_{n\to\infty} x_n + \mu \mathop{\mathrm{LIM}}_{n\to\infty} y_n$;

(ii) $\displaystyle \liminf_{n\to\infty} x_n \leqslant \mathop{\mathrm{LIM}}_{n\to\infty} x_n \leqslant \limsup_{n\to\infty} x_n$;

(iii) $\displaystyle \mathop{\mathrm{LIM}}_{n\to\infty} x_{n+1} = \mathop{\mathrm{LIM}}_{n\to\infty} x_n$.

Let $m \in \mathbb{N}$ be fixed and define $x_n = n/m - \lfloor n/m \rfloor$. What is $\mathrm{LIM}\, x_n$?

20. Show that if in Theorem 13 we drop the condition on A then the assertion is no longer true. [Hint. Let X be a vector space with basis $e_1, e_2, \ldots$, and let $A = -B = \left\{ \sum_{i=1}^{n} t_2, x_2, : t_n > 0, n = 1, 2, \ldots \right\}$. Check that $f(A) = f(B) = \mathbb{R}$ for all $f \in X'$, $f \neq 0$.]

Notes

The Hahn–Banach theorem for real normed spaces was proved by H. Hahn, *Über lineare Gleichungen in linearen Räumen*, J. für die reine und angewandte Mathematik, **157** (1927), 214–29; and by S. Banach, *Sur les fonctionnelles linéaires II*, Studia Math., **1** (1929), 223–39. The complex version, namely Theorem 6, was proved by H. F. Bohnenblust and A. Sobczyk, *Extensions of functionals on complex linear spaces*, Bull. Amer. Math. Soc., **44** (1938), 91–3.

Many dual spaces were first identified by F. Riesz, in *Sur les opérations fonctionnelles linéaires*, Comptes Rendus, **149** (1909), 974–7, and in *Untersuchungen über Systeme integrierbarer Funktionen*, Mathematische Annalen, **69** (1910), 449–97.

4. FINITE-DIMENSIONAL NORMED SPACES

As the next cautious step in our exploration of normed spaces and operators on them, we look at the 'smallest' normed spaces, namely the finite-dimensional ones. As far as the crude classification of norms is concerned, these spaces are very simple indeed: any two norms on a finite-dimensional vector space are equivalent. This can be proved in many different ways; the proof we give here is based on a lemma about the space l_1^n.

Lemma 1. The closed unit ball of l_1^n is compact.

Proof. We shall show that $B(l_1^n)$ is sequentially compact. Let $(e_i)_1^n$ be the standard basis of l_1^n, so that

$$\left\| \sum_{i=1}^n \lambda_i e_i \right\| = \sum_{i=1}^n |\lambda_i|.$$

Given $(x_k)_1^\infty \subset B = B(l_1^n)$ with $x_k = \sum_{i=1}^n \lambda_i^{(k)} e_i$, we have $|\lambda_i^{(k)}| \leq 1$ for all i and k. As a bounded sequence of complex numbers has a convergent subsequence, by repeatedly selecting subsequences, we can find a subsequence $(x_{n_k})_1^\infty$ such that $\lambda_i^{(n_k)}$ converges to some scalar λ_i for every i ($1 \leq i \leq n$).

Setting $x = \sum_{i=1}^n \lambda_i e_i$ we find that

$$\lim_{k \to \infty} \|x_{n_k} - x\| = \lim_{k \to \infty} \sum_{i=1}^n |\lambda_i^{(n_k)} - \lambda_i| = 0$$

and so $x_{n_k} \to x$. Since B is closed, $x \in B$. $\qquad \square$

Theorem 2. On a finite-dimensional vector space any two norms are equivalent.

Proof. Let V be an n-dimensional vector space with basis $(e_i)_1^n$. Let $\|\cdot\|_1$ be the l_1-norm on V given by

$$\left\| \sum_{i=1}^n \lambda_i e_i \right\|_1 = \sum_{i=1}^n |\lambda_i|$$

and let $\|\cdot\|$ be an arbitrary norm on V. It suffices to show that $\|\cdot\|$ and $\|\cdot\|_1$ are equivalent.

Let $X_1 = (V, \|\cdot\|_1)$, $S_1 = S(X_1)$ and let $f: S_1 \to \mathbb{R}$ be defined by $f(x) = \|x\|$. The set S_1 is a closed subset of the compact set $B(X_1)$ and therefore it is compact. Furthermore

$$|f(x) - f(y)| \leq \|x - y\| = \left\| \sum_{i=1}^n x_i e_i - \sum_{i=1}^n y_i e_i \right\|$$

$$\leq \sum_{i=1}^n |x_i - y_i| \|e_i\|$$

$$\leq (\max_{1 \leq i \leq n} \|e_i\|) \sum_{i=1}^n |x_i - y_i|$$

$$= (\max_{1 \leq i \leq n} \|e_i\|) |x - y\|_1$$

so f is a continuous function on the compact set S_1. Hence f attains its infimum m and supremum M on S_1. Since $f(x) = \|x\| > 0$ for all $x \in S_1$, we have $m > 0$. By the definition of f, for any $x \in V$ we have

$$m\|x\|_1 \leq \|x\| \leq M\|x\|_1. \qquad \square$$

This theorem has several easy but important consequences.

Corollary 3. Let X and Y be normed spaces, with X finite-dimensional. Then every linear operator $T: X \to Y$ is continuous. In particular, every linear functional on X is continuous.

Proof. Note that $\|x\|' = \|x\| + \|Tx\|$ is a norm on X; since $\|\cdot\|$ and $\|\cdot\|'$ are equivalent, there is an N such that $\|x\|' \leq N\|x\|$ for all x and so $\|T\| \leq N$. $\qquad \square$

Corollary 4. Any two finite-dimensional spaces of the same dimension are isomorphic.

Proof. If $\dim X = \dim Y$ then there is an invertible operator $T \in \mathscr{L}(X, Y)$. As both T and T^{-1} are bounded, X and Y are isomorphic. $\qquad \square$

Corollary 5. Every finite-dimensional space is complete.

Proof. If a space is complete in one norm then it is complete in every equivalent norm. Since, for example, the space l_∞^n is complete, the assertion follows. □

Corollary 6. In a finite-dimensional space a set is compact iff it is closed and bounded. In particular, the closed unit ball and the unit sphere are compact.

Proof. Recall that $B(l_1^n)$ is compact, and that a closed subset of a compact set is compact. □

Corollary 7. Every finite-dimensional subspace of a normed space is closed and complete.

Proof. The assertion is immediate from Corollary 5. □

In fact, as proved by Frédéric Riesz, the compactness of the unit ball characterises finite-dimensional normed spaces. We shall prove this by making use of the following variant of a lemma also due to Riesz.

Theorem 8. Let Y be a proper subspace of a normed space X.
 (a) If Y is closed then for every $\epsilon > 0$ there is a point $x \in S(X)$ whose distance from Y is at least $1 - \epsilon$:
$$d(x, Y) = \inf\{\|x - y\| : y \in Y\} \geq 1 - \epsilon.$$
 (b) If Y is finite-dimensional then there is a point $x \in S(X)$ whose distance from Y is 1.

Proof. Let $z \in X \backslash Y$ and set $Z = \mathrm{lin}\{Y, z\}$. Define a linear functional $f : Z \to \mathbb{R}$ by $f(y + \lambda z) = \lambda$ for $y \in Y$ and $\lambda \in \mathbb{R}$. Then f is a bounded linear functional since $\mathrm{Ker}\, f = Y$ is a closed subspace of Z. By the Hahn–Banach theorem, f has an extension to a bounded linear functional $F \in X^*$ with $\|F\| = \|f\| > 0$. Note that $Y \subset \mathrm{Ker}\, F$.
 (a) Let $x \in S(X)$ be such that
$$F(x) \geq (1 - \epsilon)\|F\|.$$
Then for $y \in Y$ we have
$$\|x - y\| \geq \frac{F(x - y)}{\|F\|} = \frac{F(x)}{\|F\|} \geq 1 - \epsilon.$$

(b) If Y is finite-dimensional then F attains its supremum on the compact set $S(X)$ so there is an $x \in S(X)$ such that

$$F(x) = \|F\|.$$

But then for $y \in Y$ we have

$$\|x - y\| \geqslant \frac{F(x - y)}{\|F\|} = \frac{F(x)}{\|F\|} = 1. \qquad \square$$

Corollary 9. Let $X_1 \subset X_2 \subset \cdots$ be finite-dimensional subspaces of a normed space, with all inclusions proper. Then there are unit vectors $x_1, x_2, \ldots$ such that $x_n \in X_n$ and $d(x_n, X_{n-1}) = 1$ for all $n \geqslant 2$.

In particular, an infinite-dimensional normed space contains an infinite sequence (x_n) of 1-separated unit vectors (i.e. with $\|x_n - x_m\| \geqslant 1$ for $n \neq m$).

Proof. To find $x_n \in X_n$, apply Theorem 8(b) to the pair $(X, Y) = (X_n, X_{n-1})$. $\qquad \square$

Theorem 10. A normed space is finite-dimensional if and only if its unit ball is compact.

Proof. From Corollary 6, all we have to show is that if X is infinite-dimensional then its unit ball $B(X)$ is not compact. To see this, simply take an infinite sequence $x_1, x_2, \ldots \in B(X)$ whose existence is guaranteed by Corollary 9: $\|x_i - x_j\| \geqslant 1$ for $i \neq j$. As this sequence has no convergent subsequence, $B(X)$ is not compact. $\qquad \square$

Theorem 10 is often used to prove that a space under consideration is finite-dimensional: the compactness of the unit ball tells us precisely this, without giving any information about the dimension of the space.

The above proof of Theorem 10 is based on the existence of a sequence of unit vectors $(x_n)_1^\infty$ such that $\|x_i - x_j\| \geqslant 1$ for all $i \neq j$. Let us show that, in fact, we can do better: we can make sure that the inequalities are strict. All we need is the compactness of the unit ball of a finite-dimensional normed space.

Lemma 11. Let $x_1, \ldots, x_{n-1}$ be linearly independent vectors in a real normed space X of dimension $n \geqslant 2$. Then there is a vector $x_n \in S(X)$ such that $\|x_i - x_n\| > 1$ for all i $(1 \leqslant i < n)$.

Proof. We may assume that $\dim X = n$. Let $f \in S(X^*)$ be such that $f(x_i) = 0$ for $1 \leqslant i < n$. (In other words, we require

$$K(f) = \text{lin}\{x_1,\ldots,x_{n-1}\}$$

and so we have precisely two choices for f: a functional and its negative.) Furthermore, let $g \in X^*$ be such that $g(x_i) = 1$ for every i $(1 \leqslant i < n)$. Since $S(X)$ is compact, so is

$$K = \{x \in S(X): f(x) = 1\} \neq \varnothing.$$

Let $x_n \in K$ be such that $g(x_n) = \min\{g(x): x \in K\}$. Then for $1 \leqslant i < n$ we have $\|x_n - x_i\| \geqslant f(x_n - x_i) = 1$. Since $g(x_n - x_i) = g(x_n) - 1 < g(x_n)$, the choice of x_n tells us that $x_n - x_i \notin K$, so that $\|x_n - x_i\| \neq 1$. Hence $\|x_n - x_i\| > 1$ for every i $(1 \leqslant i < n)$. $\qquad\square$

Theorem 12. Let $X_1 \subset X_2 \subset \cdots$ be subspaces of a real normed space, with $\dim X_n = n$. Then there is a sequence $x_1, x_2, \ldots$ of unit vectors such that $\|x_i - x_j\| > 1$ if $i \neq j$ and $\text{lin}\{x_1,\ldots,x_n\} = X_n$ for $n = 1, 2, \ldots$. $\qquad\square$

There is another elegant way of finding 1-separated sequences of unit vectors. This time we shall rely on the finite-dimensional form of the Hahn–Banach theorem. Let us choose vectors $x_1, x_2, \ldots \in S(X)$ and support functionals $x_1^*, x_2^*, \ldots \in S(X^*)$ as follows. Pick $x_1 \in S(X_1)$ and let $x_1^* \in S(X^*)$ be a support functional at $x_1: x_1^*(x_1) = 1$. Suppose $k < n$, $\dim X_n = n$, $x_1,\ldots,x_k \in S(X_n)$, $x_1^*,\ldots,x_k^* \in S(X_n^*)$, $x_i^*(x_i) = 1$ and $x_i^*(x_j) = 0$ for $i < j$. Then

$$W = \bigcap_{i=1}^{k} K(x_i^*) = \bigcap_{i=1}^{k} \text{Ker}\, x_i^*$$

has dimension at least $n - k$ (in fact, precisely $n - k$) and so we can pick a vector $x_{k+1} \in W \cap S(X_n)$. Let x_{k+1}^* be a support functional at x_{k+1}, i.e. let $x_{k+1}^* \in S(X_n^*)$ and $x_{k+1}^*(x_{k+1}) = 1$.

This implies that if $X_1 \subset X_2 \subset \cdots$ are subspaces of X ($\dim X_n = n$) then there are sequences $(x_n)_1^\infty \subset S(X)$ and $(x_n^*)_1^\infty \subset S(X^*)$ such that $\text{lin}\{x_1,\ldots,x_n\} = X_n$, $x_n^*(x_n) = 1$ and $x_n^*(x_m) = 0$ for $n < m$. In particular, $\|x_n - x_m\| \geqslant 1$ for $n \neq m$.

The canonical bases of l_p and $l_q = l_p^*$, where $p^{-1} + q^{-1} = 1$, are examples of pairs of sequences of this type. In fact, with $e_n = (0,\ldots,0,1,0,\ldots) \in l_p$ and $e_n^* = (0,\ldots,0,1,0,\ldots) \in l_q$, where the 1's occur in the nth places, we find that $\|e_n\| = \|e_n^*\| = e_n^*(e_n) = 1$ and $e_n^*(e_m) = 0$ for *all* $n \neq m$, not only when $n < m$. A pair of sequences satisfying these conditions is called a *normalised biorthogonal* system. To be precise, given a normed space X, a *biorthogonal system on X*

consists of vectors $x_1, \ldots, x_n$ and bounded linear functionals $x_1^*, \ldots, x_n^*$ such that $x_i^*(x_j) = \delta_{ij}$ (i.e. $x_i^*(x_j) = 0$ if $i \neq j$ and $x_i^*(x_i) = 1$). Such a system is *normalised* if $\|x_i\| = \|x_i^*\| = 1$ for all i.

Theorem 13. Let X be an n-dimensional real normed space. Then there is a normalised biorthogonal system $(x_i)_1^n, (x_i^*)_1^n$ on X.

Proof. Let $X = (\mathbb{R}^n, \|\cdot\|)$ and $S = S(X)$. For $u_1, \ldots, u_n \in \mathbb{R}^n$, write $v(u_1, \ldots, u_n)$ for the n-dimensional Euclidean volume of the n-simplex with vertices $0, u_1, \ldots, u_n$. Let $x_1, \ldots, x_n \in S$ be such that $v(x_1, \ldots, x_n)$ is maximal.

We claim that these vectors will do. What are the functionals $(x_i^*)_1^n$? Clearly x_i^* has to be defined by $x_i^*(x_i) = 1$ and $x_i^*(x_j) = 0$ for $j \neq i$. But does x_i^* have norm 1? Indeed it does, since $\|x_i^*\| > 1$ implies that $x_i^*(y_i) > 1$ for some $y_i \in S$ and so

$$v(x_1, \ldots, x_{i-1}, y_i, x_{i+1}, \ldots, x_n) > v(x_1, \ldots, x_n). \qquad \square$$

The sequence $(x_i)_1^n$ whose existence is guaranteed by Theorem 13 is called an *Auerbach system*.

In order to see what the existence of an Auerbach system really means, it is worthwhile to reformulate Theorem 13 in geometric terms. Let $x_1, \ldots, x_n$ be a basis of X. Define two norms on X: the l_1 norm and the l_∞ norm defined by taking $x_1, \ldots, x_n$ as canonical basis. Thus for $x = \sum_{i=1}^n \lambda_i x_i$ set

$$\|x\|_1 = \sum_{i=1}^n |\lambda_i| \qquad \text{and} \qquad \|x\|_\infty = \max_{1 \leq i \leq n} |\lambda_i|.$$

Let X_1 and X_∞ be the normed spaces defined by these norms. Then Theorem 13 claims precisely that $x_1, \ldots, x_n$ can be chosen in such a way that

$$B(X_1) \subset B(X) \subset B(X_\infty). \qquad (1)$$

Clearly $B(X_1)$ is the convex hull of the $2n$ vectors $\pm x_1, \pm x_2, \ldots, \pm x_n$ and $B(X_\infty)$ is the n-dimensional parallelepiped whose 2^n vertices are $\sum_{i=1}^n \epsilon_i x_i$ ($\epsilon_i \in \{-1, 1\}$).

Let us return to the opening statement of this chapter: the isomorphic classification of finite-dimensional normed spaces is trivial, with two spaces being isomorphic if and only if they have the same dimension. Based on this, one could come to the hasty verdict that there is nothing to finite-dimensional normed spaces: they are not worth studying. As it happens, this would not only be a hasty verdict but it would also be utterly incorrect. There are a great many important and interesting questions, only the isomorphic classification is not one of them. All these questions, many of which are still open, concern the metric properties of the finite-dimensional normed spaces.

Perhaps the most fundamental question is the following: given two n-dimensional normed spaces, how close are they to being isometric? Let us formulate this question precisely.

Let X and Y be isomorphic normed spaces, i.e. let X and Y be such that there is a bounded linear operator $T \in \mathscr{B}(X, Y)$ which has a bounded inverse. The *Banach–Mazur distance* between X and Y is

$$d(X, Y) = \inf\{\|T\|\|T^{-1}\| : T \in \mathscr{B}(X, Y),\ T^{-1} \in \mathscr{B}(Y, X)\}.$$

If X and Y are not isomorphic then one defines their Banach–Mazur distance to be ∞.

Note that $d(X, Y) \geqslant 1$, $d(Y, X) = d(X, Y) = d(X^*, Y^*)$ for any two spaces, and

$$d(X, Z) \leqslant d(X, Y)\, d(Y, Z)$$

for any three spaces. Thus in some sense it would be more natural to measure the 'distance' between X and Y by $\log d(X, Y)$.

What does $d(X, Y) < d$ mean in geometric terms? It means precisely that there is an invertible linear operator $T \in \mathscr{B}(X, Y)$ such that

$$\frac{1}{c_1} B(Y) \subset TB(X) \subset c_2 B(Y)$$

for some c_1 and c_2 satisfying $c_1 c_2 < d$, in other words that

$$SB(Y) \subset B(X) \subset cSB(Y)$$

for some $S \in \mathscr{B}(Y, X)$ and $c < d$. Equivalently, we may demand that

$$B(Y) \subset TB(X) \subset cB(Y)$$

for some $c < d$. Thus the distance is less than d if after a linear transformation the unit ball of one of the spaces is sandwiched between the unit ball of the other space and c times that unit ball, where $c < d$.

Corollary 4 tells us that if $\dim X = \dim Y = n$ then X and Y are isomorphic and so $d(X, Y) < \infty$. But how large can $d(X, Y)$ be? A compactness argument implies immediately that $d(X, Y) \leqslant f(n)$ for some function $f : \mathbb{N} \to \mathbb{R}$. In fact, the following simple consequence of Theorem 13 gives a bound on $d(X, Y)$.

Corollary 14. Let $X = (V, \|\cdot\|)$ be an n-dimensional normed space. Then $d(X, l_1^n) \leq n$.

Proof. Let $(x_i)_1^n$ be an Auerbach system in X so that, in particular, $(x_i)_1^n$ is a basis of V. Define a norm $\|\cdot\|_1$ on X by setting

$$\left\| \sum_{i=1}^n \lambda_i x_i \right\| = \sum_{i=1}^n |\lambda_i|.$$

Then $X_1 = (V, \|\cdot\|_1)$ is isometric to l_1^n; so we have to show that $d(X, X_1) \leq n$.

We claim that the formal identity $J : X_1 \to X$ is such that $\|J\| \leq 1$ and $\|J^{-1}\| \leq n$. Indeed, for $x = \sum_{i=1}^n \lambda_i x_i$ we have

$$\|J(x)\| = \left\| \sum_{i=1}^n \lambda_i x_i \right\| \leq \sum_{i=1}^n |\lambda_i| \|x_i\| = \sum_{i=1}^n |\lambda_i| = \|x\|_1,$$

and so $\|J\| \leq 1$. (As $\|Jx_1\| = \|x_1\|_1 = 1$, in fact $\|J\| = 1$.)

In order to estimate $\|J^{-1}\|$, let $(x_i^*)_1^n$ be the other half of the normalised biothogonal system $(x_i)_1^n$, $(x_i^*)_1^n$; in other words, let x_i^* be defined by $x_i^*(x_j) = \delta_{ij}$. Given $x = \sum_{i=1}^n \lambda_i x_i$, choose $\epsilon_i \in \{-1, 1\}$ such that $\epsilon_i \lambda_i = |\lambda_i|$ for $i = 1, \ldots, n$. Set $f = \sum_{i=1}^n \epsilon_i x_i^* \in X^*$. Then $\|f\| \leq n$ and

$$f(x) = \sum_{i=1}^n |\lambda_i| = \|x\|_1,$$

and so

$$\|J^{-1}(x)\|_1 = \|x\|_1 = f(x) \leq n \|x\|,$$

implying $\|J^{-1}\| \leq n$. $\square$

The alert reader may have noticed that the formal proof above is, in fact, unnecessary, because the result is a trivial consequence of (1), the relation equivalent to Theorem 13. Indeed, $B(X_\infty) \subset nB(X_1)$ and so $B(X_1) \subset B(X) \subset nB(X_1)$. Thus $d(X, X_1) \leq n$. But X_1 is isometric to l_1^n, so that $d(X, l_1^n) \leq n$.

An immediate consequence of Corollary 14 is that $d(X, Y) \leq n^2$ for any two n-dimensional spaces, but this trivial estimate is far from being best possible. In 1948 Fritz John proved an essentially best-possible upper bound for $d(X, Y)$ by first bounding the distance of an n-dimensional space from l_2^n, rather than from l_1^n. Before giving this result, let us think a little about the Banach–Mazur distance from a Euclidean space.

An easy compactness argument implies that if $\dim X = \dim Y = n$ then $d(X, Y)$ is attained: there is an operator $S \in \mathscr{B}(Y, X)$, say, such that

$$SB(Y) \subset B(X) \subset dSB(Y)$$

where $d = d(X, Y)$. Now if $Y = l_2^n = (\mathbb{R}^n, \|\cdot\|)$ then $B(Y)$ is the Euclidean unit ball

$$D = \left\{(x_i)_1^n \in \mathbb{R}^n : \sum_{i=1}^n x_i^2 \leq 1\right\}.$$

For a linear operator $S \in \mathscr{L}(\mathbb{R}^n, \mathbb{R}^n)$, the image SD of the Euclidean ball D is an ellipsoid centred at 0; conversely, for every ellipsoid $E \subset \mathbb{R}^n$, with centre at 0, there is a linear map $S \in \mathscr{L}(\mathbb{R}^n, \mathbb{R}^n)$ such that $SD = E$. Thus $d(X, l_2^n) \leq d$ if and only if there is an ellipsoid E, with centre 0, such that

$$E \subset B(X) \subset dE.$$

The following elegant and important theorem of John shows that one can obtain a good upper bound for $d(X, l_2^n)$ be taking the ellipsoid E of minimal volume containing $B(X)$.

Theorem 15. (John's theorem) Let $X = (\mathbb{R}^n, \|\cdot\|)$ be a normed space with unit ball $B = B(X)$. Then there is a unique ellipsoid D of minimal (Euclidean) volume containing B. Furthermore,

$$n^{-1/2}D \subset B \subset D.$$

In particular, $d(X, l_2^n) \leq n^{1/2}$.

Proof. A simple compactness argument shows that the infimum of the volumes of ellipsoids containing B is attained. Furthermore, every ellipsoid of minimal volume is centred at 0.

Let us first prove uniqueness. Suppose that D and D' are ellipsoids of minimal volume containing B. If $T \in \mathscr{L}(\mathbb{R}^n)$ is invertible then $T(D)$ and $T(D')$ are ellipsoids of minimal volume containing $T(B)$. Hence we may assume that

$$D = \left\{x \in \mathbb{R}^n : \sum_{i=1}^n x_i^2 \leq 1\right\} \quad \text{and} \quad D' = \left\{x \in \mathbb{R}^n : \sum_{i=1}^n \frac{x_i^2}{a_i^2} \leq 1\right\},$$

where $a_i > 0 \ (i = 1, \ldots, n)$.

Denoting by v_n the volume of D, we have $\text{vol}\, D' = v_n \prod_{i=1}^{n} a_i$, and so $\prod_{i=1}^{n} a_i = 1$. Let E be the ellipsoid

$$E = \left\{ x \in \mathbb{R}^n : \sum_{i=1}^{n} \tfrac{1}{2} x_i^2 \left(1 + \frac{1}{a_i^2} \right) \leq 1 \right\}.$$

Then

$$B \subset D \cap D' \subset E$$

and

$$\frac{(\text{vol}\, E)^2}{(\text{vol}\, D)^2} = \prod_{i=1}^{n} \frac{2}{1 + 1/a_i^2} = \prod_{i=1}^{n} \frac{2a_i^2}{1 + a_i^2} = \prod_{i=1}^{n} \frac{2a_i}{1 + a_i^2} < 1$$

since $\prod_{i=1}^{n} a_i = 1$ and not every a_i is 1. This contradicts the assumption that D was an ellipsoid of minimal volume containing B.

Let us turn to the proof of $n^{-1/2} D \subset B$. Suppose that this is not the case, so that B has a boundary point in the interior of $n^{-1/2} D$. By taking a support plane of B at such a point and rotating B to make this support plane parallel to the plane of the axes $x_2, x_3, \ldots, x_n$, we may assume that

$$B \subset P = \left\{ x \in \mathbb{R}^n : |x_1| \leq \frac{1}{c} \right\}$$

for some $c > n^{1/2}$.

For $a > b > 0$ define an ellipsoid $E_{a,b}$ by

$$E_{a,b} = \left\{ x \in \mathbb{R}^n : a^2 x_1^2 + b^2 \sum_{i=2}^{n} x_i^2 \leq 1 \right\},$$

so that $(\text{vol}\, D)/(\text{vol}\, E_{a,b}) = ab^{n-1}$. If $x \in B$ then $x \in D \cap P$ and so

$$a^2 x_1^2 + b^2 \sum_{i=2}^{n} x_i^2 = (a^2 - b^2) x_1^2 + b^2 \sum_{i=1}^{n} x_i^2$$

$$\leq \frac{a^2 - b^2}{c^2} + b^2.$$

Hence $B \subset E_{a,b}$ and $\text{vol}\, E_{a,b} < \text{vol}\, D$ whenever

$$\frac{a^2 - b^2}{c^2} + b^2 \leq 1 \qquad \text{and} \qquad ab^{n-1} > 1.$$

Thus, to complete the proof, it suffices to show that these inequalities are satisfied for some choice of $0 < b < a$. This is a trivial matter which, unfortunately, looks a little untidy.

Put $0 < \epsilon < \frac{1}{2}$, $a = (1 + \epsilon + 2\epsilon^2)^{n-1}$ and $b = 1 - \epsilon$. Then $ab^{n-1} > 1$. Also,

$$
\frac{a^2 - b^2}{c^2} + b^2 = \frac{(1 + \epsilon + 2\epsilon^2)^{2(n-1)}}{c^2} + (1 - \epsilon)^2 \left(1 - \frac{1}{c^2}\right)
$$

$$
= 1 + 2\epsilon \left(\frac{n-1}{c^2} - 1 + \frac{1}{c^2}\right) + O(\epsilon^2)
$$

$$
= 1 + 2\epsilon \left(\frac{n}{c^2} - 1\right) + O(\epsilon^2) < 1
$$

if $\epsilon > 0$ is sufficiently small. $\square$

Theorem 15 can be reformulated in terms of inscribed ellipsoids: there is a unique ellipsoid D of *maximal* volume contained in B and $D \subset B \subset n^{1/2}D$. Furthermore, if B is not a unit ball but only a bounded convex body in $\mathbb{R}^n$ then some translate B' of B satisfies $D \subset B' \subset nD$ for the ellipsoid D of maximal volume contained in B' (see Exercise 18).

Corollary 16. Let X and Y be n-dimensional normed spaces. Then $d(X, Y) \leq n$.

Proof. $d(X, Y) \leq d(X, l_2^n) d(l_2^n, Y) \leq n^{1/2} n^{1/2} = n$. $\square$

Although the estimate in Corollary 16 does seem somewhat crude, since the Banach–Mazur distance between X and Y is estimated by going through l_2^n, it is, in fact, close to being best possible: in 1981 the Russian mathematician E. D. Gluskin proved that there is a constant $c > 0$ such that for every n there are n-dimensional normed spaces X and Y with $d(X, Y) \geq cn$. At the time this result was extremely surprising; the proof was based on a probabilistic argument which has become an important tool in the so-called *local theory of Banach spaces*, the theory concerning finite-dimensional spaces.

Exercises

1. Let $x_1, \ldots, x_n$ be non-zero elements in a normed space. For $\lambda = (\lambda_i)_1^n \in \mathbb{R}^n$ set

$$
\|\lambda\|_0 = \max \left\{ \left\| \sum_{i=1}^n \epsilon_i \lambda_i x_i \right\| : \epsilon_i \in \{-1, 1\} \ (i = 1, \ldots, n) \right\}.
$$

Show that $\|\cdot\|_0$ is a norm on $\mathbb{R}^n$. Show also that for $\lambda = (\lambda_i)_{i=1}^n$ and $j = (1)_{i=1}^n \in \mathbb{R}^n$ we have

$$\|\lambda\|_0 = \|j\|_0 \|\lambda\|_\infty = \|j\|_0 \max_i |\lambda_i|.$$

2. Let $X_i = (V_i, \|\cdot\|)$ $(i = 1,\ldots,n)$ be normed spaces. For $1 \leqslant p \leqslant \infty$ and $x = (x_i)_1^n \in V = V_1 \oplus \cdots \oplus V_n$ set

$$\|x\|_p = \|(\|x_i\|)_1^n\|_p.$$

Show that $\|\cdot\|_p$ is a norm on V. Show also that any two of these norms are equivalent. [The space $(V, \|\cdot\|_p)$ is usually denoted by $X_1 \oplus_p X_2 \oplus_p \cdots \oplus_p X_n$; it is a direct sum of $X_1,\ldots,X_n$.]

3. Let $|\cdot|$ be a symmetric norm on $\mathbb{R}^n$, i.e.

$$\left| \sum_{i=1}^n \lambda_i e_i \right| = \left| \sum_{i=1}^n |\lambda_i| e_i \right|, \quad \text{with } |e_i| = |(0,\ldots,0,1,0,\ldots,0)| = 1$$

and let $X_1,\ldots,X_n$ be normed spaces. Show that

$$\|(x_1,x_2,\ldots,x_n)\| = |(\|x_1\|_1, \|x_2\|_2, \ldots, \|x_n\|_n)|$$

is a norm on $X = X_1 \oplus \cdots \oplus X_n$, inducing the product topology on this space. Check that each X_i is a closed subspace of X and that X is complete iff each X_i is complete.

4. Let $x_1,\ldots,x_n$ be unit vectors in a normed space such that

$$\left\| \sum_{i=1}^n \lambda_i x_i \right\| \leqslant c^2 \max_{1 \leqslant i \leqslant n} |\lambda_i|$$

for all $\lambda_i,\ldots,\lambda_n \in \mathbb{R}$. Show that for $k^2 \leqslant n$ there are unit vectors

$$u_i = \sum_{j \in U_i} \mu_j x_j \quad (i = 1,\ldots,k)$$

such that $U_i \cap U_{i'} = \varnothing$ for $i \neq i'$ and

$$\left\| \sum_{i=1}^k \lambda_i u_i \right\| \leqslant c \max_{1 \leqslant i \leqslant k} |\lambda_i|$$

for all $\lambda_1,\ldots,\lambda_k \in \mathbb{R}$.

5[+]. Let X be a finite-dimensional real normed space, $x_0 \in X$ and $T \in \mathscr{B}(X)$. Can $\{T^n x_0\}_1^\infty$ be dense in X? [HINT: First prove it for complex normed spaces.]

6[+]. Let Y be a proper subspace of a finite-dimensional normed space X. Can one always find a vector $z_0 \in X$ $(z_0 \neq 0)$ such that

$\|y+z_0\| \geqslant \|y\|$ for all $y \in Y$? Or a subspace $Z \subset X$ such that $Y + Z = X$ and $\|y+z\| \geqslant \|y\|$ for all $y \in Y$ and $z \in Z$?

7. Let Y be a finite codimensional closed subspace of a normed space X and let $T \in \mathcal{L}(X, Z)$, where Z is a normed space. Show that if the restriction of T to Y is continuous then so is T.

8. Let $x_1, \ldots, x_n$ be unit vectors in a normed space. Let $0 < \epsilon < \frac{1}{2}$ and suppose that

$$\left\| \sum_{i=1}^{n} \lambda_i x_i \right\| \leqslant (1+\epsilon) \max_{1 \leqslant i \leqslant n} |\lambda_i|$$

for all $\lambda_1, \ldots, \lambda_n \in \mathbb{R}$. Prove that

$$\left\| \sum_{i=1}^{n} \lambda_i x_i \right\| \geqslant (1-\epsilon) \max_{1 \leqslant i \leqslant n} |\lambda_i|$$

for all $\lambda_1, \ldots, \lambda_n \in \mathbb{R}$.

9. Let X and Y be n-dimensional normed spaces. Show that the Banach–Mazur distance between X and Y is attained, i.e. there is an operator $T_0 \in \mathcal{B}(X, Y)$ such that $T_0^{-1} \in \mathcal{B}(Y, X)$ and

$$d(X, Y) = \inf\{\|T\|\|T^{-1}\| : T \in \mathcal{B}(X, Y)\} = \|T_0\|\|T_0^{-1}\|.$$

10. Show that if $2 \leqslant p \leqslant \infty$ then

$$d(l_p^n, l_2^n) \leqslant n^{\frac{1}{2} - 1/p}.$$

11. Let X be a normed space with unit ball $B = B(X)$ and suppose that B can be covered by a finite number of translates of $\frac{1}{2}B$: there are $x_1, \ldots, x_n \in X$ such that

$$B \subset \bigcup_{i=1}^{n} (x_i + \tfrac{1}{2}B) = \bigcup_{i=1}^{n} B(x_i, \tfrac{1}{2}).$$

By making use of Corollary 7 but no subsequent result, prove that X is finite-dimensional, and deduce Theorem 10.

12. Let Y be a finite-dimensional subspace of an infinite-dimensional normed space X. Show that for every $\epsilon > 0$ there is an $x \in S(X)$ such that $\|y\| \leqslant (1+\epsilon)\|x+y\|$ for every $y \in Y$.

13. A sequence $(x_n)_1^\infty$ in a normed space is said to be a *basic sequence* if $x_n \neq 0$ for every n, and there is a constant K such that

$$\left\| \sum_{k=1}^{m} \lambda_k x_k \right\| \leqslant K \left\| \sum_{k=1}^{n} \lambda_k x_k \right\|$$

for all $1 \leqslant m \leqslant n$ and scalars $\lambda_1, \ldots, \lambda_n$. The minimal K satisfying this is the *basis constant* of $(x_n)_1^\infty$. Deduce from the result in the previous exercise that every infinite-dimensional normed space contains a basic sequence.

14. Let $(x_n)_1^\infty$ be a basic sequence in a Banach space. Show that $(x_n)_1^\infty$ is a Schauder basis of its closed linear span, i.e. every $x \in \overline{\text{lin}}(x_n)_1^\infty$ has a unique representation in the form $x = \sum_{k=1}^\infty \lambda_k x_k$, with the series convergent in norm.

15. Deduce from Corollary 9 that a Banach space cannot have a countably infinite algebraic basis, i.e. if $X = \text{lin}(e_n)_1^\infty$, where $e_1, e_2, \ldots$ are independent, then X is incomplete.

16. Let X and Y be finite-dimensional normed spaces and set

$$N_K = \{T \in B(X, Y) : \|T\|, \|T^{-1}\| \leqslant K\}.$$

Prove that N_K is a compact subset of $B(X, Y)$ (with the operator norm).

17[+]. Let X be an infinite-dimensional separable Banach space such that for every subspace $Y \subset X$ ($\dim Y = n$) there is an operator $T_n \in B(Y, l_2^n)$ such that $\|T_n\|\|T_n^{-1}\| \leqslant K$. Prove that there is an operator $T \in B(X, l_2)$ such that $\|T\|\|T^{-1}\| \leqslant K$.

18. Let $B \subset \mathbb{R}^n$ be a bounded convex body (i.e. $\text{Int}\, B \neq \varnothing$). Show that there is an ellipsoid D of centre 0 such that some translate B' of B satisfies $D \subset B' \subset nD$.

19. Prove that if a linear functional is continuous on a closed finite-codimensional subspace then it is a bounded linear functional.

20. Let $X_1 = (V_1, \|\cdot\|)$ and $X_2 = (V_2, \|\cdot\|)$ be subspaces of a normed space $X = (V, \|\cdot\|)$, with X_1 finite-dimensional. Show that if $V = V_1 \oplus V_2$ then X is the direct sum of X_1 and X_2, i.e. the projections $p_{X_i} : X = X_1 + X_2 \to X_i$ ($i = 1, 2$) are continuous.

Notes

Fritz John's theorem was proved in the paper *Extremum problems with inequalities as subsidiary conditions*, in *Courant Anniversary Volume*, Interscience, New York, 1948, pp. 187–204. The original paper of E. D. Gluskin, *The diameter of the Minkowski compactum is roughly equal to* n, Funct. Anal. Appl., **15** (1981), 72–3, is an all-too-concise account of his celebrated results.

Excellent accounts of much of the excitement concerning finite-dimensional spaces can be found in the following volumes: V. D. Milman and G. Schechtman, *Asymptotic Theory of Finite Dimensional*

Normed Spaces, Lecture Notes in Mathematics, vol. 1200, Springer-Verlag, Berlin, Heidelberg, 1986, viii + 156 pp., G. Pisier, *Volume of Convex Bodies and Geometry of Banach Spaces*, Cambridge University Press, 1989, xv + 250 pp., and N. Tomczak-Jaegermann, *Banach–Mazur Distance and Finite Dimensional Operator Ideals*, Pitman, 1989.

5. THE BAIRE CATEGORY THEOREM AND THE CLOSED-GRAPH THEOREM

In this chapter we shall present several of the most fundamental and useful results in classical functional analysis: the Baire category theorem, the Banach–Steinhaus theorem, the theorem of the condensation of singularities, the open-mapping theorem and the closed-graph theorem. All these results are very closely related; indeed, they are practically variants of each other. Nevertheless, it is rather useful to emphasize the many facets of the same phenomenon, because even a subtle variation in the formulation can make quite a difference in an application.

The simplest member of this group of results is due to Baire and concerns metric spaces. Recall that a set Y in a topological space X is *dense in* X if $\bar{Y}$, the closure of Y, is X. In other words, Y is dense if $Y \cap G \neq \varnothing$ whenever G is a non-empty open set. We shall give several equivalent formulations of the Baire category theorem: here is the first.

Theorem 1. Let $G_1, G_2, \ldots$ be a sequence of dense open subsets of a complete metric space X. Then $G = \bigcap_{n=1}^{\infty} G_n$ is dense in X.

Proof. As in a normed space, for $x \in X$ and $r > 0$ we set

$$D_r(x) = \{y \in X : d(x,y) < r\} \qquad \text{and} \qquad B_r(x) = \{y \in X : d(x,y) \leq r\}.$$

Although $\overline{D_r(x)}$ need not be $B_r(x)$, we have $\overline{D_r(x)} \subset B_r(x) \subset D_{r+\epsilon}(x)$ for all $\epsilon > 0$. Since G_n is a dense open subset of X, for all $x \in X$ and $r > 0$ the open ball $D_r(x)$ meets G_n in a non-empty open set so there are $y \in X$ and $s > 0$ such that $B_s(y) \subset D_{2s}(y) \subset G_n \cap D_r(x)$. This is the property we shall exploit.

Let then $x \in X$ and $r > 0$. We have to show that $D_r(x) \cap G \neq \varnothing$. Let us construct sequences of points $x_1, x_2, \ldots \in X$ and positive numbers $r_1, r_2, \ldots$ as follows. Choose $x_1 \in X$ and $0 < r_1 < 1$ such that

$B_{r_1}(x_1) \subset G_1 \cap D_r(x)$. Next choose $x_2 \in X$ and $0 < r_2 < \frac{1}{2}$ such that $B_{r_2}(x_2) \subset G_2 \cap D_{r_1}(x_1)$, then choose $x_3 \in X$ and $0 < r_3 < \frac{1}{3}$ such that $B_{r_3}(x_3) \subset G_3 \cap D_{r_2}(x_2)$, etc. By construction, $r_n \to 0$ and $B_{r_1}(x_1) \supset B_{r_2}(x_2) \supset \cdots$ and so, by the completeness of X, $\bigcap_{n=1}^{\infty} B_{r_n}(x_n) = \{x_0\}$ for some $x_0 \in X$. As $x_0 \in B_{r_1}(x_1) \subset D_r(x)$ and $x_0 \in B_{r_n}(x_n) \subset G_n$ for all n, we have $x_0 \in D_r(x) \cap G$. $\qquad\square$

Because of its importance and many applications, we give some other forms of Theorem 1 and thereby explain the name of the result as well. The first is a slightly weaker but very useful variant.

Theorem 1'. If a complete metric space is the union of countably many closed sets then at least one of the closed sets has non-empty interior.

Proof. Let $X = \bigcup_{n+1}^{\infty} F_n$, where X is a complete metric space and each F_n is closed. Setting $G_n = X \backslash F_n$ we find that $\bigcap_{n=1}^{\infty} G_n = \varnothing$ so that at least one of the open sets G_n is not dense in X: $\bar{G}_n \neq X$. But $\text{Int } F_n = X \backslash \bar{G}_n$. $\qquad\square$

Strictly speaking, Theorem 1' says precisely that the set G in Theorem 1 is not empty. However, replacing X by $B_r(x)$, we see that $G \cap B_r(x)$ is not empty either, and so G is dense. Thus Theorem 1 is an easy consequence of Theorem 1'.

A subset Y of a topological space X is said to be *nowhere dense in X* if the closure of Y has empty interior: $\text{Int } \bar{Y} = \varnothing$. Note that a set Y is nowhere dense if and only if its closure, $\bar{Y}$, is nowhere dense. A subset $Z \subset X$ is *meagre in X* or *of the first category in X* if it is the union of countably many nowhere-dense sets. Clearly, the union of countably many meagre sets is meagre. A subset of X is *of the second category* in X if it is not meagre in X. Thus U is of the second category in X if, whenever $F_1, F_2, \ldots$ are closed sets and $U \subset \bigcup_{n=1}^{\infty} F_n$, then $\text{Int } F_n \neq \varnothing$ for some n.

Let us use this terminology to give another reformulation of the Baire category theorem.

Theorem 1''. The complement of a meagre subset of a complete metric space is dense. In particular, a complete metric space is of the second category.

Also, in a complete metric space the complement of a set of the first category (a meagre set) is of the second category.

Proof. Suppose $Y = \bigcup_{n=1}^{\infty} Y_n \subset X$ with X complete and each Y_n nowhere dense in X. Then $\bar{Y}_n$ is nowhere dense in X so $G_n = X \backslash \bar{Y}_n$ is a dense open set and therefore $G = \bigcap_{n=1}^{\infty} G_n$ is dense. Clearly $G \subset X \backslash Y$.

To see the third assertion, note that the union of two meagre sets is meagre. □

The two parts of the last variant are trivially equivalent since if A is a set of the second category in a topological space X and $B \subset X$ is meagre then $A \backslash B$ is again of the second category.

The intuitive meaning of a set of the second category is that it is a *large* set. Thus, saying that 'the set of points having a certain property (e.g. where a given function is continuous) is of the second category' is rather similar to saying that '*almost all* points have this property', i.e. the set of points not having this property has measure zero. Of course, the second statement makes sense only if we have a measure on the space, while being of the second category is an intrinsic property of a metric space. For example, given a collection $\mathscr{C}$ of convex bodies in $\mathbb{R}^n$, loosely speaking one may say that 'almost all sets in $\mathscr{C}$ have property P', meaning that in the natural (Hausdorff) metric on $\mathscr{C}$, the set of convex bodies having P is of the second category.

It is perhaps worth pointing out that the two notions are only similar but not comparable, even if we restrict our attention to $[0, 1] \subset \mathbb{R}$. Indeed, $[0, \frac{1}{2}] \subset [0, 1]$ does not have full measure, yet is of the second category, and one can construct a subset of $[0, 1]$ with measure 1 that is of the first category (cf. Exercise 23).

In 1897 Osgood proved the following pretty result about continuous functions. Let $f_1, f_2, \ldots : [0, 1] \to \mathbb{R}$ be continuous functions such that for each $t \in [0, 1]$ the sequence $(f_n(t))_1^{\infty}$ is bounded. Then there is an interval $[a, b] \subset [0, 1]$ $(a < b)$ on which the sequence $f_1, f_2, \ldots$ is *uniformly* bounded.

Let us see that the Baire category theorem implies the following extension of this result.

Theorem 2. (Principle of uniform boundedness) Let U be a set of the second category in a metric space X and let $\mathscr{F}$ be a family of continuous functions $f : X \to \mathbb{R}$ such that $\{f(u) : f \in \mathscr{F}\}$ is bounded for every $u \in U$. Then the elements of $\mathscr{F}$ are uniformly bounded in some ball $B_r(x_0)$ $(x_0 \in X; r > 0)$, i.e. $|f(x)| \leq n$ holds for some n and all $f \in \mathscr{F}$ and $x \in B_r(x_0)$). In particular, the conclusion holds if X is complete and $\{f(x) : f \in \mathscr{F}\}$ is bounded for every $x \in X$.

Proof. For $n \geqslant 1$ let

$$F_n = \{x \in X : |f(x)| \leqslant n \text{ for all } f \in \mathscr{F}\} = \bigcap_{f \in \mathscr{F}} f^{-1}([-n, n]).$$

Then F_n, as the intersection of a family of closed sets, is closed. By our assumption,

$$U \subset \bigcup_{n=1}^{\infty} F_n,$$

so by the definition of a set of the second category, $\text{Int} F_n \neq \varnothing$ for some n.

The second assertion is immediate from Theorem 1″. □

The next result is only a little more than a reformulation of Theorem 2 in the setting of linear maps.

Theorem 3. (Banach–Steinhaus theorem) Let X and Y be normed spaces, let U be a set of the second category in X and let $\mathscr{F} \subset \mathscr{B}(X, Y)$ be a family of bounded linear operators such that

$$\sup\{\|Tu\| : T \in \mathscr{F}\} < \infty$$

for all $u \in U$. Then $\|T\| \leqslant N$ for some N and all $T \in \mathscr{F}$. In particular, the conclusion holds if X is complete and $\sup\{\|Tx\| : T \in \mathscr{F}\} < \infty$ for all $x \in X$.

Proof. For $T \in \mathscr{F}$ the function $X \to \mathbb{R}$, given by $x \mapsto \|Tx\|$, is continuous. Hence, by Theorem 2, there is a ball $B_r(x_0)$ $(x_0 \in X; r > 0)$ such that $\|Tx\| \leqslant n$ for some n and all $T \in \mathscr{F}$ and $x \in B_r(x_0)$. As in the proof of Theorem 2.2, this implies that $\|T\| \leqslant N = n/r$ for all $T \in \mathscr{F}$.

Indeed, let $T \in \mathscr{F}$ and $x \in B(X)$. Then $x_0 + rx$ and $x_0 - rx$ are in $B_r(x_0)$ and so

$$\|Tx\| = \frac{1}{2r}\|T(x_0 + rx - \{x_0 - rx\})\| \leqslant \frac{n}{r} = N$$

and therefore $\|T\| \leqslant N$ for all $T \in \mathscr{F}$. □

The last result has the following immediate consequence concerning the condensation of singularities.

Theorem 4. Let X and Y be normed spaces, with X complete, and let $T_{nm} \in \mathscr{B}(X, Y)$ $(n, m = 1, 2 \ldots)$ be such that

$$\varlimsup_{m \to \infty} \|T_{nm}\| = \infty \qquad \text{for all } n.$$

Then there is a set $U \subset X$ of the second category in X such that for $u \in U$ we have

$$\overline{\lim_{m \to \infty}} \, \|T_{nm}(u)\| = \infty \qquad \text{for all } n.$$

Proof. For a fixed n, let $V_n \subset X$ be the set of vectors v such that $\overline{\lim}_{m \to \infty} \|T_{nm}(v)\| < \infty$. Then, by Theorem 3, V_n is of the first category. Hence so is $V = \bigcup_{n=1}^{\infty} V_n$ and thus by Theorem 1″ the set $U = X \backslash V$ is of the second category in X. □

The next major result, the open-mapping theorem, is a consequence of the Baire category theorem and the important lemma below. Recall that for a normed space X we write $D(X)$ for the open unit ball and $B(X)$ for the closed unit ball; furthermore, we use the notation $D_r(X) = rD(X)$ and $B_r(X) = rB(X)$.

Lemma 5. Suppose X and Y are normed spaces, X is complete, $T \in \mathcal{B}(X, Y)$ and $\overline{T(D_r(X))} \supset D_s(Y)$. Then $T(D_r(X)) \supset D_s(Y)$.

Proof. We may and shall assume that $r = s = 1$. Let $A = D(Y) \cap T(D(X))$; then $\bar{A} = B(Y)$. Given $z \in D(Y)$, choose δ such that $\|z\| < 1 - \delta < 1$ and set $y = z/(1-\delta)$. We shall show that $y \in T(D(X))/(1-\delta)$ and so $z \in T(D(X))$.

Let us define a sequence $(y_i)_0^{\infty} \subset Y$. Set $y_0 = 0$ and then choose successively $y_1, y_2, \ldots \in Y$ such that $y_n - y_{n-1} \in \delta^{n-1}A$ and $\|y_n - y\| < \delta^n$. Having chosen $y_0, \ldots, y_{n-1}$, we can find a suitable y_n since $y \in D(y_{n-1}, \delta^{n-1})$ and the set $\delta^{n-1}A$ is dense in $D_{\delta^{n-1}}(Y)$.

By the definition of A, there exists a sequence $(x_n)_1^{\infty} \subset X$ such that $Tx_n = y_n - y_{n-1}$ and $\|x_n\| < \delta^{n-1}$. Putting

$$x = \sum_{n=1}^{\infty} x_n$$

we have

$$\|x\| \leqslant \sum_{n=1}^{\infty} \|x_n\| < (1-\delta)^{-1} \qquad \text{and} \qquad Tx = \sum_{n=1}^{\infty} (y_n - y_{n-1}) = y.$$

Consequently $y \in T(D(X))/(1-\delta)$ and so $z \in T(D(X))$. As z was an arbitrary point of $D(Y)$, this implies that $D(Y) \subset TD(X)$, as claimed. □

Theorem 6. (Open-mapping theorem) Let X and Y be Banach spaces and let $T \in \mathcal{B}(X, Y)$ be a surjection: $T(X) = Y$. Then T is an open map, i.e. if $U \subset X$ is open then so is TU.

Proof. Set $G = T(D(X))$. Since T is a linear map, it suffices to show that G contains a neighbourhood of 0 in Y.

Note that $T(D_r(X)) = rG$ and the closure of rG is $r\overline{G}$. Since $Y = T(X) = \bigcup_{n=1}^{\infty} n\overline{G}$, by the Baire category theorem we see that $\operatorname{Int} n\overline{G} \neq \varnothing$ for some n and so $\operatorname{Int}\overline{G} \neq \varnothing$. The set $\overline{G}$ is convex and symmetric about 0 so $\operatorname{Int} \overline{G} \neq \varnothing$ implies that $\overline{G} \supset D_r(Y)$ for some $r > 0$. Indeed, if $D(x_0, r) \subset \overline{G}$ then $D(-x_0, r) \subset \overline{G}$ and so the convexity of $\overline{G}$ implies that $D(0, r) = D_r(Y) \subset \overline{G}$. But then, by Lemma 5, $G = T(D(X)) \supset D_r(Y)$. $\square$

The final two results are easy but very useful consequences of the open-mapping theorem.

Theorem 7. (Inverse-mapping theorem) If T is a one-to-one bounded linear operator from a Banach space X onto a Banach space Y then T^{-1} is also bounded.

Proof. Since T is a bijection, the inverse T^{-1} exists and belongs to $\mathscr{L}(Y, X)$. By Theorem 6, for some $r > 0$ we have $T(D(X)) \supset D_r(Y)$. Thus $\|T^{-1}\| \leqslant 1/r$. $\square$

Theorem 8. (Closed-graph theorem) Let X and Y be Banach spaces and $T \in L(X, Y)$. Then T is bounded iff its graph,

$$\Gamma(T) = \{(x, Tx) : x \in X\} \subset X \times Y$$

is closed in $X \times Y$ in the product topology.

Proof. One of the implications is trivial: the graph of a continuous map into a Hausdorff space is closed, so if T is bounded, $\Gamma(T)$ is closed.

Suppose then that $\Gamma(T)$ is closed. Let Z be the direct sum of X and Y endowed with the norm $\|(x, y)\| = \|x\| + \|y\|$. This norm induces the product topology on $X \times Y$ and so, by assumption, $\Gamma(T)$ is a closed subset of Z. In fact, Z is easily seen to be complete, so $\Gamma(T)$ is a closed subspace of the Banach space Z and therefore it is itself a Banach space. The linear map $U : \Gamma(T) \to X$, given by $U((x, y)) = x$, is a norm-decreasing bijection and so, by Theorem 7, the inverse operator U^{-1} is bounded. But then $\|T\| \leqslant \|U^{-1}\|$ since

$$\|Tx\| \leqslant \|x\| + \|Tx\| = \|U^{-1}(x)\| \leqslant \|U^{-1}\|\|x\|.   \square$$

It is worth pointing out that the closed-graph theorem often makes it much easier to prove that an operator is continuous. When is a linear operator $T : X \to Y$ continuous? If $x_1, x_2, \ldots \in X$ and $x_n \to x_0$ imply that $Tx_n \to Tx_0$. Now if X and Y are Banach spaces, the closed-graph theorem tells us that in order to prove the continuity of a linear operator $T: X \to Y$, it suffices to show that if $x_n \to x_0$ and $Tx_n \to y_0$ then $Tx_n \to Tx_0$, i.e. $Tx_0 = y_0$ – a considerable gain!

Exercises

1. Let $f: \mathbb{R}^+ \to \mathbb{R}$ be a continuous function such that $\lim_{n\to\infty} f(nx) = 0$ for all $x > 0$. Show that $\lim_{x\to\infty} f(x) = 0$.

2+. Let $f: \mathbb{R} \to \mathbb{R}$ be infinitely differentiable. Suppose for every $x \in \mathbb{R}$ there is a k_x such that $f^{(k)}(x) = 0$ for all $k \geqslant k_x$. Prove that f is a polynomial.

3. Let $K = [0,1]$, $X = C^{\mathbb{R}}(K)$ and, for $n \geqslant 1$, set $F_n =$
$$\left\{ f \in X : \exists\, t \in K \text{ such that } \left| \frac{f(t+h) - f(t)}{h} \right| \leqslant n\; \forall\, h \text{ with } t+h \in K \right\}.$$

   Prove that F_n is closed and nowhere dense in $C^{\mathbb{R}}(K)$. Deduce that the set of continuous nowhere-differentiable real-valued functions on $[0,1]$ is dense in $C^{\mathbb{R}}(K)$.

4. Prove that if a vector space is a Banach space with respect to two norms then the topologies induced by the norms are either equivalent or incomparable (i.e. neither is stronger than the other).

5. Let V be a vector space with algebraic basis $e_1, e_2, \ldots$ (so $\dim V = \infty$ and every $v \in V$ is a unique linear combination of the e_i) and let $\|\cdot\|$ be a norm on V. Show that $(V, \|\cdot\|)$ is incomplete.

6. Let X be a normed space and $S \subset X$. Show that if $\{f(x) : x \in S\}$ is bounded for every linear functional $f \in X^*$ then the set S is bounded: $\|x\| \leqslant K$ for some K and all $x \in S$. (Using fancy terminology that will become clear in Chapter 8, a *weakly bounded* set is norm bounded.)

7. Deduce from the result in the previous exercise that if two norms on a vector space V are not equivalent then there is a linear functional $f \in V'$ which is continuous in one of the norms and discontinuous in the other.

8. Let X be a closed subspace of $L_1(0,2)$. Suppose for every $f \in L_1(0,1)$ there is an $F \in X$ whose restriction to $(0,1)$ is f. Show that there is a constant c such that our function F can always be chosen to satisfy $\|F\| \leqslant c\|f\|$.

9. Let $1 \leqslant p, q \leqslant \infty$ and let $A = (a_{ij})_1^\infty$ be a scalar matrix. Suppose for every $x = (x_j)_1^\infty \in l_p$ the series $\sum_{j=1}^\infty a_{ij} x_j$ is convergent for

every i, and $y = (y_i)_1^\infty \in l_q$, where $y_i = \sum_{j=1}^\infty a_{ij}x_j$. Show that the map $A: l_p \to l_q$, defined by $x \mapsto y$, is a bounded linear map.

10. Let $\varphi_n : [0,1] \to \mathbb{R}^+$ $(n = 1,2,\ldots)$ be uniformly bounded continuous functions such that

$$\int_0^1 \varphi_n(x)\, dx \geq c$$

for some $c > 0$. Suppose $c_n \geq 0$, $(n = 1,2,\ldots)$ and

$$\sum_{n=1}^\infty c_n \varphi_n(x) < \infty$$

for every $x \in [0,1]$. Prove that

$$\sum_{n=1}^\infty c_n < \infty.$$

11. For two sequences of scalars $x = (x_i)_1^\infty$ and $y = (y_i)_1^\infty$ set

$$(x,y) = \sum_{i=1}^\infty x_i y_i.$$

Let $1 < p, q < \infty$ be conjugate indices and let $x^{(n)} = (x_i^{(n)})_{i=1}^\infty \in l_p$ $(n = 1,2,\ldots)$. Show that $(x^{(n)}, y) \to 0$ for every $y \in l_q$ iff $(x^{(n)})_1^\infty$ is norm bounded (i.e. $\|x^{(n)}\| \leq K$ for some K and all n) and $x_i^{(n)} \to 0$ for every i. (This is a characterization of sequences in l_p *tending weakly* to 0; cf. Exercise 6.)

12. Let $x^{(n)} = (x_i^{(n)})_{i=1}^\infty \in l_1$ $(n = 1,2,\ldots)$. Show that $(x^{(n)}, y) \to 0$ for every $y \in c_0$ iff $(x^{(n)})_1^\infty$ is bounded and $x_i^{(n)} \to 0$ for every i. (Similarly to Exercise 11, this is a characterization of sequences in l_1 tending weakly to 0.)

13. Show that for $1 \leq p < \infty$ the L_p-norm

$$\|f\|_p = \left(\int_0^1 |f(t)|^p\, dt \right)^{1/p}$$

on $C[0,1]$ is dominated by the uniform norm $\|f\| = \sup_{0 \leq t \leq 1} |f(t)|$ and deduce that $C[0,1]$ is incomplete in the norm $\|\cdot\|_p$.

14. Let P be a *projection* on a Banach space X, i.e. let P be a linear map of X into itself such that $P^2 = P$. Show that X is the direct sum of the subspaces $\operatorname{Ker} P = P^{-1}(0)$ and $\operatorname{Im} P = PX$ if and only if P is bounded.

15. Let X and Y be normed spaces and $T \in \mathcal{L}(X,Y)$. Show that the graph of T is closed iff whenever $x_n \to 0$ and $Tx_n \to y$ then $y = 0$.

As we shall see in the following exercises, the results of this chapter are very useful in proving the fundamental properties of Schauder bases. Recall from Chapter 2 that a sequence $(x_n)_1^\infty$ is a *Schauder basis* or simply a *basis* of a Banach space X if every $x \in X$ has a unique representation in the form $\sum_{i=1}^\infty \lambda_i x_i$, with the series convergent in norm.

16. Let $(x_n)_1^\infty$ be a basis of a Banach space X. Define $P_n \in \mathcal{L}(X)$ by

$$P_n\left(\sum_{i=1}^\infty \lambda_i x_i\right) = \sum_{i=1}^n \lambda_i x_i,$$

and for $x \in X$ set $|\!|\!|x|\!|\!| = \sup_n \|P_n x\|$. Prove that $|\!|\!|\cdot|\!|\!|$ is a norm on X and X that is complete in this norm as well.

17. Let X, $(x_n)_1^\infty$ and $(P_n)_1^\infty$ be as in the previous exercise. Prove that $P_n \in \mathcal{B}(X)$ and $\sup_n\|P_n\| < \infty$. [The number $\sup_n\|P_n\|$ is called the *basis constant* of $(x_n)_1^\infty$; cf. Exercise 4.13.]

18. Let $x_1, x_2, \ldots$ be non-zero vectors in a Banach space X, with $\overline{\operatorname{lin}}(x_n)_1^\infty = X$. Prove that $(x_n)_1^\infty$ is a basis of X if and only if there is a constant K such that

$$\left\|\sum_{i=1}^m \lambda_i x_i\right\| \leq K\left\|\sum_{i=1}^n \lambda_i x_i\right\|$$

for all $1 \leq m < n$ and scalars $\lambda_1, \ldots, \lambda_n$.

19. Define a sequence of functions $\chi_n : [0,1] \to [-1,1]$ as follows. Set $\chi_1(t) \equiv 1$ and for $n \geq 2$ define $k \geq 0$ by $2^k < n \leq 2^{k+1}$. Set $l = n - 2^k - 1$ and define

$$\chi_n(t) = \begin{cases} 1 & \text{if } l \leq 2^k x < l + \tfrac{1}{2}, \\ -1 & \text{if } l + \tfrac{1}{2} \leq 2^k x < l + 1, \\ 0 & \text{otherwise.} \end{cases}$$

The sequence $(\chi_n(t))_1^\infty$ is called the *Haar system*. Prove that the Haar system is a basis in $L_p(0,1)$ for $1 \leq p < \infty$.

20. Define a sequence of continuous functions $\varphi_n : [0,1] \to [0,1]$ as follows. Set $\varphi_1(t) \equiv 1$ and for $n \geq 2$ put

$$\varphi_n(t) = \int_0^t \chi_{n-1}(u)\, du,$$

where χ_k is the kth Haar function, defined in the previous exercise. The sequence $(\varphi_n(t))_1^\infty$ is called the *Schauder system*. Prove that the Schauder system is a basis in $C^{\mathbb{R}}[0,1]$ with the uniform norm. Show also that the basis constant is 1.

21. Let $(x_n)_1^\infty$ be a sequence in a normed space such that

$$\sum_{n=1}^\infty |f(x_n)| < \infty$$

for every $f \in X^*$. Show that there is a constant M such that

$$\sum_{n=1}^\infty |f(x_n)| \le M\|f\|$$

for every $f \in X^*$.

22. Use the Baire category theorem to deduce the result in Exercise 4.15: the linear span (*not* the *closed* linear span!) of an infinite sequence of linearly independent vectors in a normed space cannot be complete.

23+. Construct a set $S \subset [0, 1]$ of measure 1 that is of the first category in $[0, 1]$.

Notes

Most of the results in this chapter can be found in Stefan Banach's classic, cited in Chapter 2. The original paper of R. Baire is *Sur la convergence de variables réelles*, Annali Mat. Pura e Appl., **3** (1899), 1–122; Osgood's theorem is in W. F. Osgood, *Non-uniform convergence and the integration of series term by term*, Amer. J. Math., **19** (1897), 155–90. The Banach–Steinhaus result was proved in S. Banach and H. Steinhaus, *Sur le principe de la condensation de singularités*, Fundamenta Math., **9** (1927), 51–7.

6. CONTINUOUS FUNCTIONS ON COMPACT SPACES AND THE STONE–WEIERSTRASS THEOREM

In this chapter we shall study one of the most important classical Banach spaces, the space $C(K)$ of continuous complex-valued functions on a compact Hausdorff space K, with the uniform (supremum) norm

$$\|f\| = \sup_{x \in K} |f(x)| = \max_{x \in K} |f(x)|.$$

In fact, as we shall see in Chapter 8, every Banach space is a closed subspace of $C(K)$ for some compact Hausdorff space K, although this does not really help in the study of a general Banach space.

First we shall show that $C(K)$ is large enough, namely that it contains sufficiently many functions; for example, every continuous function on a closed subset of K can be extended to a continuous function on the whole of K without increasing its norm. This result, which is a special case of the Tietze–Urysohn extension theorem, is thus reminiscent of the Hahn–Banach theorem: it ensures that the space $C(K)$ is large, just as the Hahn–Banach theorem ensured that the dual space of a normed linear space was large. It is worth emphasizing that the existence of a good stock of continuous functions on a topological space X cannot be taken for granted, not even when X seems to be very pleasant. For example, it can happen that X is a countable Hausdorff space and still every continuous function on X is constant (see Exercise 19).

In the second part of the chapter we shall show that $C(K)$ is not too large in the sense that it is the norm closure (i.e. the closure in the topology induced by the norm) of a good many 'small' subspaces of functions.

We shall start with a standard result in analytic topology which is likely to be familiar to many readers.

A topological space is said to be *normal* if every pair of disjoint closed sets can be separated by disjoint open sets: if A and B are disjoint closed sets then there are disjoint open sets U and V such that $A \subset U$ and $B \subset V$. Equivalently, a space is normal if for all $A_0 \subset U_0$, where A_0 is closed and U_0 is open, one can find a closed set A_1 and an open set U_1 such that $A_0 \subset U_1 \subset A_1 \subset U_0$.

Lemma 1.
  (a) In a Hausdorff space any two disjoint compact sets can be separated by open sets.
  (b) Every compact Hausdorff space is normal.

Proof. (a) Let A and B be non-empty disjoint compact sets in a Hausdorff space. For $a \in A$ and $b \in B$, let $U(a,b)$ and $V(a,b)$ be disjoint open neighbourhoods of a and b. Let us fix a point $a \in A$. Clearly, $B \subset \bigcup_{b \in B} V(a,b)$ and so, as B is compact, $B \subset \bigcup_{i=1}^{n} V(a,b_i)$ for some $b_1, \ldots, b_n \in B$. Set

$$U(a) = \bigcap_{i=1}^{n} U(a,b_i) \quad \text{and} \quad V(a) = \bigcup_{i=1}^{n} V(a,b_i).$$

Then $U(a)$ and $V(a)$ are disjoint open sets, $U(a)$ is a neighbourhood of a and $V(a)$ contains B.

Now note that $\bigcup_{a \in A} U(a)$ is an open cover of A and so, as A is compact, $A \subset \bigcup_{i=1}^{m} U(a_i)$ for some $a_1, \ldots, a_m \in A$. Let

$$U = \bigcup_{i=1}^{m} U(a_i) \quad \text{and} \quad V = \bigcap_{i=1}^{m} V(a_i).$$

Then U and V are disjoint open sets, $A \subset U$ and $B \subset V$.

(b) In a compact space every closed set is compact, so the assertion follows from (a). □

If A and B are disjoint closed sets in a metric space X then there is a continuous function $f : X \rightarrow [0,1]$ such that f is 0 on A and 1 on B; indeed,

$$f(x) = \begin{cases} \min\{d(x,A)/d(x,B),1\} & x \notin B \\ 1 & x \in B \end{cases}$$

will do. The next important result tells us that such an f exists not only on metric spaces but also on normal spaces.

Theorem 2. (Urysohn's lemma) Let A and B be disjoint closed subsets of a normal space X. Then there is a continuous function $f : X \rightarrow [0,1]$ such that $A \subset f^{-1}(0)$ and $B \subset f^{-1}(1)$.

Proof. Let $q_0 = 0$, $q_1 = 1$ and let $q_2, q_3, \ldots$ be an enumeration of the rationals strictly between 0 and 1. Set $U_0 = \varnothing$, $A_0 = A$, $U_1 = X \backslash B$ and $A_1 = X$. Let us construct a sequence of closed sets $A_0, A_1, \ldots$ and a sequence of open sets $U_0, U_1, \ldots$ such that $U_i \subset A_i$ and if $q_i < q_j$ then $A_i \subset U_j$. Suppose we have found $A_0, \ldots, A_{n-1}$ and $U_0, \ldots, U_{n-1}$. What shall we choose for A_n and U_n? Let q_k be the maximal q_i $(i < n)$ which is less than q_n and let q_l be the minimal q_i $(i < n)$ which is greater than q_n. Then $A_k \subset U_l$ and so, as X is normal, one can find a closed set A_n and an open set U_n such that $A_k \subset U_n \subset A_n \subset U_l$.

Having constructed $A_0, A_1, \ldots$ and $U_0, U_1, \ldots$, it is easy to define a suitable continuous function f: for $x \in X$ set

$$f(x) = \inf\{q_n : x \in A_n\}.$$

Clearly $0 \le f(x) \le 1$ and f is 0 on A and 1 on B. All we have to check is that f is continuous. Suppose then that $0 \le q_k < f(x_0)$. Then $x_0 \notin A_k$ and so $X \backslash A_k$ is a neighbourhood of x_0; for $x \in X \backslash A_k$ we have $f(x) \ge q_k$. Similarly, if $f(x_0) < q_l \le 1$ then $x \in U_l$. So U_l is a neighbourhood of x_0 and for $x \in U_l$ we have $f(x) \le q_l$. $\qquad\square$

We are ready to present one of the main results of this section. By Urysohn's lemma, if A and B are disjoint closed (possibly empty) subsets of a normal space X then there is a continuous function $g = g(A, B) : X \to [-1, 1]$ such that $g|A = -1$ and $g|B = 1$. Equivalently, if C is a closed subset of X and h is a continuous function from C into the two-point set $\{-1, 1\}$ then h has an extension to a continuous function $g : X \to [-1, 1]$. The Tietze–Urysohn extension theorem below claims that *every* continuous function $f : C \to [-1, 1]$ has such an extension. This result is easily proved by using Urysohn's lemma to provide us with continuous functions on X whose restrictions to C are better and better approximations of f.

Indeed, setting $A_0 = \{x \in C : f(x) \le 0\}$ and $B_0 = \{x \in C : f(x) \ge \frac{1}{2}\}$, there is a continuous function $G_0 : X \to [0, \frac{1}{2}]$ such that $G_0|A_0 = 0$ and $G_0|B_0 = \frac{1}{2}$. Then $f' = f - G_0|C$ maps C into $[-1, \frac{1}{2}]$ – a very good start indeed. Next, set

$$A_1 = \{x \in C : f'(x) \le -\tfrac{1}{2}\}, \qquad B_1 = \{x \in C : f'(x) \ge 0\}$$

and take in a continuous function $H_1 : C \to [-\frac{1}{2}, 0]$ such that $H_1|A_1 = -\frac{1}{2}$ and $H_2|B_1 = 0$. Then $F_1 = G_0 + H_1$ is a good approximation of f on C, namely $f - F_1|C$ maps C into $[-\frac{1}{2}, \frac{1}{2}]$ and $\|F_1\| \le \frac{1}{2}$. Continuing in this way, we get continuous functions F_2, F_3, ... with $\|F_n\| \le 2^{-n}$ and

$f - \sum_{i=1}^{n} F_i | C$ mapping C into $[-2^{-n}, 2^{-n}]$. Thus $F = \sum_{i=1}^{\infty} F_i$ will have the required properties.

For the sake of completeness, in the proof below we shall present a streamlined version of this approach.

Theorem 3. (Tietze–Urysohn extension theorem) Let C be a closed subset of a normal space X and let $f: C \to [-1, 1]$ be a continuous function. Then f has a continuous extension to the whole space X, i.e. there is a continuous function $F: X \to [-1, 1]$ such that $F|C = f$.

Proof. We shall construct F as the uniform limit of continuous functions on the whole of X whose restrictions to C are better and better approximations of f. To be precise, set $f_0 = f$,

$$A_0 = \{x \in C: f_0(x) \leqslant -\tfrac{1}{3}\} \quad \text{and} \quad B_0 = \{x \in C: f_0(x) \geqslant \tfrac{1}{3}\}.$$

By Urysohn's lemma there is a continuous function $F_0: X \to [-\tfrac{1}{3}, \tfrac{1}{3}]$ such that $F_0|A_0 = -\tfrac{1}{3}$ and $F_0|B_0 = \tfrac{1}{3}$. Set $f_1 = f_0 - F_0|C$ and note that f_1 is a continuous map of C into $[-\tfrac{2}{3}, \tfrac{2}{3}]$. In general, having constructed a continuous function

$$f_n: C \to [-(\tfrac{2}{3})^n, (\tfrac{2}{3})^n],$$

define

$$A_n = \{x \in C: f_n(x) \leqslant -\tfrac{1}{3}(\tfrac{2}{3})^n\} \quad \text{and} \quad B_n = \{x \in C: f_n(x) \geqslant \tfrac{1}{3}(\tfrac{2}{3})^n\}.$$

By Urysohn's lemma there is a continuous function

$$F_n: X | [-\tfrac{1}{3}(\tfrac{2}{3})^n, \tfrac{1}{3}(\tfrac{2}{3})^n]$$

such that

$$F_n|A_n = -\tfrac{1}{3}(\tfrac{2}{3})^n \quad \text{and} \quad F_n|B_n = \tfrac{1}{3}(\tfrac{2}{3})^n.$$

Set

$$f_{n+1} = f_n - F_n|C$$

and note that f_{n+1} is a continuous map C into $[-(\tfrac{2}{3})^{n+1}, (\tfrac{2}{3})^{n+1}]$.

Let $f_0, f_1, \ldots$ and $F_0, F_1, \ldots$ be the sequences of functions constructed in this way. Then $|F_n(x)| \leqslant \tfrac{1}{3}(\tfrac{2}{3})^n$ for all $x \in X$ and $n \geqslant 0$ and so $F(x) = \sum_{n=0}^{\infty} F_n(x)$ is continuous, being a uniform limit of continuous functions, and

$$|F(x)| \leqslant \tfrac{1}{3} \sum_{n=0}^{\infty} (\tfrac{2}{3})^n = 1.$$

Furthermore, for $x \in C$ and $n \geqslant 0$ we have

$$\left| f(x) - \sum_{k=0}^{n} F_k(x) \right| = |f_{n+1}(x)| \leqslant (\tfrac{2}{3})^{n+1}$$

and so $f(x) = F(x)$. $\qquad\qquad\qquad\qquad\qquad\qquad\qquad\qquad\square$

It may be worth noting that it is trivial to ensure that $|F(x)| \leqslant 1$: if $\tilde{F}(x)$ is any continuous extension of f then

$$F(x) = \min\{1, \max\{-1, \tilde{F}(x)\}\}$$

will do.

From now on, let K be a compact Hausdorff space and let us consider the space $C(K)$ of all continuous (complex-valued) functions on K. This is a Banach space in the uniform norm $\|f\| = \sup\{|f(x)| : x \in K\}$. The results above tell us that $C(K)$ is rather rich in functions. Namely, $C(K)$ *separates* the points of X: for any two distinct points x and y there is a function $f \in C(K)$ such that $f(x) \neq f(y)$. Also, if A and B are disjoint closed subsets of K then for some $f \in C(K)$ we have $f(a) = 0$ and $f(b) = 1$ for all $a \in A$ and $b \in B$. Finally, if $A \subset K$ is closed and $f \in C(A)$ then $f = F|A$ for some $F \in C(K)$. To see the last statement, simply write f as a sum of its real part and its complex part.

The space $C(K)$ is closed under uniform limits, and this is another source of functions that can be guaranteed to belong to $C(K)$. This leads us to consider *relatively compact* subsets of $C(K)$, that is subsets in which every sequence has a subsequence convergent to some element of $C(K)$.

Recall that a topological space is *compact* if every open cover has a finite subcover, it is *countably compact* if every countable open cover has a finite subcover and it is *sequentially compact* if every sequence has a convergent subsequence. Given a metric space X and $N \subset X$, the set N is an ϵ-*net* if for every $x \in X$ there is a point $y \in N$ with $d(x,y) < \epsilon$. The ϵ-nets one tends to consider are finite. A metric space is *totally bounded* or *relatively compact* if it contains a finite ϵ-net for every $\epsilon > 0$, or, equivalently, if every sequence has a Cauchy subsequence. For a metric space, the properties of being compact, countably compact and sequentially compact coincide; furthermore, a metric space is compact iff it is complete and relatively compact. In particular, a subset of a complete metric space is relatively compact iff its closure is compact; a closed subset of a complete metric space is compact iff it is relatively compact.

In order to characterize the relatively compact subsets of the complete metric space $C(K)$, let us introduce some more terminology. A set $S \subset C(K)$ is *uniformly bounded* if it is bounded in the supremum norm, i.e. if $\|f\| \leq N$ for some N and all $f \in S$. We say that S is *equicontinuous at a point* $x \in K$ if for every $\epsilon > 0$ there is a neighbourhood U_x of x such that if $y \in U_x$ and $f \in S$ then $|f(x) - f(y)| < \epsilon$. Furthermore, S is *equicontinuous* if it is equicontinuous at every point.

Theorem 4. (Arzelà–Ascoli theorem) For a compact space K, a set $S \subset C(K)$ is relatively compact if and only if it is uniformly bounded and equicontinuous.

Proof. (a) Suppose that S is a totally bounded subset of $C(K)$. Let $x \in K$ and $\epsilon > 0$. The set S contains a finite ϵ-net: there is a finite set $\{f_1, \ldots, f_n\} \subset S$ such that if $f \in S$ then $\|f - f_i\| < \epsilon$. Then, trivially, $\|f\| < \epsilon + \max_{1 \leq i \leq n} \|f_i\|$ and so S is uniformly bounded. Furthermore, as f_i is continuous, for every $x \in K$ there is a neighbourhood U_i of x such that $|f_i(x) - f_i(y)| < \epsilon$ for $y \in U_i$. Then $U = \bigcap_{i=1}^{n} U_i$ is a neighbourhood of x showing that S is equicontinuous at x. Indeed, for $f \in S$ let i $(1 \leq i \leq n)$ be such that $\|f - f_i\| < \epsilon$. Then for $y \in U$ we have

$$|f(x) - f(y)| \leq |f(x) - f_i(x)| + |f_i(x) - f_i(y)| + |f_i(y) - f(y)| < 3\epsilon.$$

(b) Now suppose that S is uniformly bounded and equicontinuous. Given $\epsilon > 0$, for $x \in K$ let U_x be an open neighbourhood of x such that $|f(x) - f(y)| < \epsilon$ if $f \in S$ and $y \in U_x$. Then $K = \bigcup_{x \in K} U_x$; let $U_{x_1}, \ldots, U_{x_n}$ be a finite subcover of K. Since S is uniformly bounded, the set $R = \{(f(x_1), \ldots, f(x_n)) : f \in S\}$ is a bounded subset of l_∞^n, the vector space $\mathbb{C}^n$ endowed with the supremum norm. (As all norms on a finite-dimensional vector space are equivalent, we could have chosen any norm on $\mathbb{C}^n$.) A bounded set in l_∞^n is totally bounded because a bounded closed set in l_∞^n is compact; therefore R is totally bounded so there are functions $f_1, \ldots, f_m \in S$ such that the set

$$\{(f_i(x_1), f_i(x_2), \ldots, f_i(x_n)) : 1 \leq i \leq m\}$$

is an ϵ-net in R.

We claim that the functions $f_1, \ldots, f_m$ form a 3ϵ-net in S. Indeed, given $f \in S$, there is an i $(1 \leq i \leq m)$ such that $|f(x_j) - f_i(x_j)| < \epsilon$ for *all* j $(1 \leq j \leq n)$. Let now $x \in K$ be an arbitrary point. Then $x \in U_{x_j}$ for some j $(1 \leq j \leq n)$. Consequently,

$$|f(x) - f_i(x)| \leq |f(x) - f(x_j)| + |f(x_j) - f_i(x_j)| + |f_i(x_j) - f_i(x)| < 3\epsilon. \quad \square$$

In conclusion, let us see that $C^{\mathbb{R}}(K)$, the space of continuous real-valued functions on K, is closed under monotone pointwise limits. This is Dini's theorem.

Theorem 5. Let K be a compact topological space and let $f_0, f_1, f_2, \ldots \in C^{\mathbb{R}}(K)$. Suppose that for every $x \in K$ the sequence $(f_n(x))_1^{\infty}$ is monotone increasing and tends to $f_0(x)$. Then $(f_n)_1^{\infty}$ converges uniformly to f_0, i.e. $\|f_n - f_0\| \to 0$.

Proof. Let $\epsilon > 0$. For every $x \in K$ there is a natural number n_x such that $f_{n_x}(x) > f_0(x) - \epsilon$. As $f_0 - f_{n_x}$ is a non-negative continuous function, x has an open neighbourhood U_x such that if $y \in U_x$ then

$$0 \leqslant f_0(y) - f_{n_x}(y) < \epsilon.$$

Note that if $n \geqslant n_x$ and $y \in U_x$ then

$$0 \leqslant f_0(y) - f_n(y) \leqslant f_0(y) - f_{n_x}(y) < \epsilon.$$

Since $\bigcup_{x \in K} U_x$ is certainly a cover of K and K is compact, K is covered by a finite number of these sets: $K = \bigcup_{i=1}^{k} U_{x_i}$ for some points $x_1, \ldots, x_k$. The last inequality implies that if $n \geqslant n_0 = \max\{n_{x_1}, \ldots, n_{x_k}\}$ then $|f_0(y) - f_n(y)| < \epsilon$ for all y, and so $\|f_0 - f_n\| \leqslant \epsilon$. $\square$

It is worth noting that all three conditions are needed in Theorem 5, namely that K is compact, the sequence $(f_n(x))_1^{\infty}$ of continuous functions is pointwise increasing and the pointwise limit, f_0 is a continuous function (see Exercise 13).

Now let us turn to one of the most fundamental results in functional analysis, the Stone–Weierstrass theorem, telling us that the sapce $C(K)$ is not too large: it is the uniform closure of a good many pleasant subspaces, more precisely, it is the uniform closure of a good many subalgebras. Recall that a topological space is said to be *locally compact* if every point has a compact neighbourhood. For a locally compact Hausdorff space L, let

$C(L) = \{f \in \mathbb{C}^L : f \text{ is continuous and bounded}\},$

$C_0(L) = \{f \in C(L) : \text{the set } \{x : |f(x)| \geqslant \epsilon\} \text{ is compact for every } \epsilon > 0\}$

and

$C_c(L) = \{f \in C(L) : \text{supp} f \text{ is compact}\},$

where supp f, the *support* of a (not necessarily continuous) function f, is the closure of the set $\{x : f(x) \neq 0\}$.

Note that if L is compact then $C(L)$ is just the set of all continuous functions; similarly, $C_c(L)$ is the set of all continuous functions with compact support and $C_0(L)$ is the set of all continuous function f on L for which $\{x \in L : |f(x)| \geq \epsilon\}$ is compact for every $\epsilon > 0$.

If f and g are functions on the same space, define λf, $f + g$ and fg by pointwise operations. With these operations, $C(L)$ is a *commutative algebra*. This algebra has a unit, the identically 1 function. By definition, $C_c(L) \subset C_0(L) \subset C(L)$ and $C_c(L)$ and $C_0(L)$ are subalgebras of $C(L)$. We know that the vector space $C(L)$ is a Banach space with the *uniform (supremum) norm*:

$$\|f\| = \sup\{|f(x)| : x \in L\}.$$

Theorem 6. Let L be a locally compact Hausdorff space. The function $\|\cdot\|$ is a norm on $C(L)$ and the space $C(L)$ is complete in this norm; $C_0(L)$ is a closed subspace of $C(L)$ and $C_c(L)$ is a dense subspace of $C_0(L)$. Furthermore, for $f, g \in C(L)$ one has

$$\|fg\| \leq \|f\|\|g\| \tag{1}$$

and 1, the identically 1 function, has norm 1.

Proof. The only assertion needing proof is that $C_c(L)$ is dense in $C_0(L)$. Given $f \in C_0(L)$ and $\epsilon > 0$, set $K_0 = \{x \in L : |f(x)| \geq \epsilon\}$. Then K_0 is compact. For $x \in K_0$ let U_x be an open neighbourhood of x with compact closure $\bar{U}_x$. Then $K_0 \subset \bigcup_{x \in K_0} U_x$ and so $K_0 \subset \bigcup_{i=1}^n U_{x_i}$ for some $x_1, \ldots, x_n \in K_0$. Setting $K_1 = \bigcup_{i=1}^n \bar{U}_{x_i}$, we find that K_1 is a compact set such that $K_0 \subset \operatorname{Int} K_1$.

By Urysohn's lemma (Theorem 2) there is a continuous function $h : K_1 \to [0, 1]$ that is 1 on K_0 and 0 on $K_1 \backslash \operatorname{Int} K_1$. Extend h to a function g on L by setting it to be 0 outside K_1:

$$g(x) = \begin{cases} h(x) & \text{if } x \in K_1, \\ 0 & \text{if } x \in L \backslash K_1. \end{cases}$$

Then $g \in C_c(L)$ and so $gf \in C_c(L)$. Clearly $\|gf - f\| \leq \epsilon$. □

An algebra A which is also a normed space and whose norm satisfies (1) is called a *normed algebra*. If the algebra has an identity e and the norm of e is 1 then it is called a *unital normed algebra*. If the norm is complete then the algebra is a *Banach algebra*. We saw in Chapter 2 that $\mathscr{B}(X)$, the set of bounded linear operators on a normed space X, is a unital normed algebra and if X is complete then $\mathscr{B}(X)$ is a unital

Banach algebra. The trivial part of Theorem 6 shows that $C(L)$ is a commutative unital Banach algebra, $C_0(L)$ is a commutative Banach algebra and $C_c(L)$ is a commutative normed algebra.

The spaces $C^{\mathbf{R}}(L)$, $C_0^{\mathbf{R}}(L)$ and $C_c^{\mathbf{R}}(L)$ are defined analogously; they consist of the appropriate continuous *real-valued* functions. These spaces are not only real normed algebras but they are also *lattices* under the natural lattice operations $f \vee g$ and $f \wedge g$.

Given real-valued functions f and g on a set S, define new functions $f \vee g$ and $f \wedge g$ on S by setting, for $x \in S$,

$$(f \vee g)(x) = \max\{f(x), g(x)\} \quad \text{and} \quad (f \wedge g)(x) = \min\{f(x), g(x)\}.$$

We call $f \vee g$ the *join of f and g* and $f \wedge g$ the *meet of f and g*. The join and the meet are the *lattice operations*.

Of course, the join of two functions is just their maximum and the meet is just their minimum. It is clear that $C^{\mathbf{R}}(L)$, $C_0^{\mathbf{R}}(L)$ and $C_c^{\mathbf{R}}(L)$ are all closed under the lattice operations.

For a set A of bounded functions on a set S, the *uniform closure $\bar{A}$* of A is the closure of A in the uniform topology on $\mathcal{F}_b(S)$, the space of bounded functions on S with the uniform norm (Example 2.1 (ii)). Thus a function f on S belongs to $\bar{A}$ if for every $\epsilon > 0$ there exists a $g \in A$ such that $|f(x) - g(x)| < \epsilon$ for all $x \in S$.

In this chapter we shall study the uniformly closed subalgebras of $C(L)$, $C_0(L)$, $C^{\mathbf{R}}(L)$ and $C_0^{\mathbf{R}}(L)$; so we are particularly interested in uniform approximations by elements from various subsets of these algebras. As the next result shows, the lattice operations are very useful when we look for such approximations.

Theorem 7. Let K be a compact Hausdorff space and let A be a subset of $C^{\mathbf{R}}(K)$ which is closed under the lattice operations. Then $\bar{A}$, the uniform closure of A, is precisely the set of those continuous (real-valued) functions on K that can be arbitrarily approximated at every pair of points by functions from A.

Remark. Note that A is not assumed to be a subalgebra, not even a subspace, it is merely a *set* closed under the lattice operations, i.e. it is a *sublattice* of $C^{\mathbf{R}}(K)$.

Proof. It is obvious that the uniform closure $\bar{A}$ is at most as large as claimed.

Suppose then that $f \in C^{\mathbf{R}}(K)$ can be approximated by elements of A at every pair of points. We have to show that $f \in \bar{A}$.

Let $\epsilon > 0$. The existence of approximations of f means precisely that for $x, y \in K$ there exists a function $f_{x,y} \in A$ such that

$$|f(x) - f_{x,y}(x)| < \epsilon \quad \text{and} \quad |f(y) - f_{x,y}(y)| < \epsilon.$$

Put $U_{x,y} = \{z \in K : f_{x,y}(z) < f(z) + \epsilon\}$. Then $U_{x,y}$ is an open set containing x and y. Fixing x, we find that $K = \bigcup_{y \in K} U_{x,y}$, and so a finite number of the sets $U_{x,y}$ also cover K, say $K = \bigcup_{i=1}^{n} U_{x,y_i}$. By assumption, the minimum of the corresponding functions $f_{x,y_1}, \ldots, f_{x,y_n}$, namely $f_x = f_{x,y_1} \wedge f_{x,y_2} \wedge \cdots \wedge f_{x,y_n}$, also belongs to A. Clearly

$$f_x(x) > f(x) - \epsilon \quad \text{and} \quad f_x(y) < f(y) + \epsilon \quad \text{for all } y \in K.$$

Let $V_x = \{y \in K : f_x(y) > f(y) - \epsilon\}$. Then V_x is open and $K = \bigcup_{x \in K} V_x$ so that, as K is compact, $K = \bigcup_{i=1}^{m} V_{x_i}$ for some $x_1, \ldots, x_m \in K$. Let $f_\epsilon = f_{x_1} \vee f_{x_2} \vee \cdots \vee f_{x_m}$ be the maximum of the corresponding functions. By assumption, $f_\epsilon \in A$, and

$$f(x) - \epsilon < f_\epsilon(x) < f(x) + \epsilon \quad \text{for all } x \in K.$$

As there is such an f_ϵ for every $\epsilon > 0$, $f \in \bar{A}$. □

Lemma 8. A uniformly closed subalgebra A of $\mathscr{F}_b(S)$, the space of bounded real-valued functions on a set S is also closed under the lattice operations.

Proof. Since $(f \vee g)(x) = \max\{f(x), g(x)\} = \frac{1}{2}\{f(x) + g(x) + |f(x) - g(x)|\}$ and $f \wedge g = f + g - f \vee g$, it suffices to show that $|f| \in A$ if $f \in A$. Furthermore, as for $n \geq 1$ we have $|nf| = n|f|$, and $f \in A$ holds iff $nf \in A$, it suffices to show that if $f \in A$ and $\|f\| = \sup_{x \in S} |f(x)| < 1$ then $|f| \in A$.

Let then $f \in A$, $\|f\| < 1$ and $\epsilon > 0$. Let $\sum_{k=0}^{\infty} c_k(t - \frac{1}{2})^k$ be the Taylor series for $(t + \epsilon^2)^{1/2}$ about $t = \frac{1}{2}$. As this series converges uniformly on $[0, 1]$, there is a polynomial

$$P(t) = \sum_{k=0}^{n} c_k(t - \tfrac{1}{2})^k$$

such that

$$|P(t) - (t + \epsilon^2)^{1/2}| < \epsilon \quad \text{for all } 0 \leq t \leq 1.$$

Hence

$$|P(s^2) - (s^2 + \epsilon^2)^{1/2}| < \epsilon \quad \text{for all } -1 \leq s \leq 1$$

and so

$$|P(s^2) - |s|| < 2\epsilon \quad \text{for all } -1 \leq s \leq 1.$$

Set $Q(s) = P(s^2) - P(0)$. Then Q is a real polynomial with constant term 0, and so $f \in A$ implies that $Q(f) \in A$. Furthermore, for $-1 \leq s \leq 1$ we have

$$|Q(s) - |s|| \leq |P(s^2) - |s|| + |P(0)| < 4\epsilon.$$

Hence

$$|Q(f)(x) - |f(x)|| < 4\epsilon \quad \text{for all } x \in K$$

and so

$$\|Q(f) - |f|\| \leq 4\epsilon,$$

showing that $|f|$ can be approximated by the function $Q(f) \in A$ within 4ϵ. Since A is uniformly closed, $|f| \in A$. $\square$

We are ready to prove the first form of the main result of this section. We say that a set A of functions on a set S *separates the points of S* if for all $x, y \in S$ $(x \neq y)$ there is a function $f \in A$ with $f(x) \neq f(y)$.

Theorem 9. (Stone–Weierstrass theorem) Let K be a compact space and let A be an algebra of real-valued continuous functions on K that separates the points of K. Then $\bar{A}$, the uniform closure of A, is either $C^{\mathbb{R}}(K)$ or the algebra of all continuous real-valued functions vanishing at a single point $x_\infty \in K$.

Proof. (a) Suppose first that there is no point $x_\infty \in K$ such that $f(x_\infty) = 0$ for all $f \in A$. It is easily checked that then for all $x \neq y$ $(x, y \in K)$ there is a function $f \in A$ such that $f(x) \neq f(y)$, $f(x) \neq 0$ and $f(y) \neq 0$. Then any function on K, say $g \in \mathbb{R}^K$, can be approximated at x and y by elements of A. In fact, we can do rather better than only approximate: since $(f(x), f(y))$ and $(f^2(x), f^2(y))$ are linearly independent vectors in $\mathbb{R}^2$, some linear combination of them is $(g(x), g(y))$. So for some linear combination $h = sf + tf^2$ of f and f^2 we have $h(x) = g(x)$ and $h(y) = g(y)$. By Theorem 7 and Lemma 8 we have $\bar{A} = C^{\mathbb{R}}(K)$.

(b) Suppose now that $f(x_\infty) = 0$ for some $x_\infty \in K$ and all $f \in A$. We have to show that if $g \in C^{\mathbb{R}}(K)$ and $g(x_\infty) = 0$ then $g \in \bar{A}$.

Let us adjoin the constant functions to A, i.e. let $B \subset C^{\mathbb{R}}(K)$ be the algebra of functions of the form $f + c$ where $f \in A$ and $c \in \mathbb{R}$. Given $\epsilon > 0$, by part (a) there exists a function $h = f + c \in B$, with $f \in A$ and

$c \in \mathbb{R}$, such that $\|f+c-g\| < \epsilon$. As $g(x_\infty) = f(x_\infty) = 0$ we have $|c| < \epsilon$ and so $\|f-g\| < 2\epsilon$. Hence $g \in \bar{A}$.   □

Let us see now what Theorem 9 tells us about subalgebras of $C(L)$. For a locally compact Hausdorff space L, the *one-point compactification* *of* L is the topological space on $L_\infty = L \bigcup \{x_\infty\}$ where $x_\infty \notin L$ and a set $S \subset L_\infty$ is open iff (a) $S \bigcap L$ is open in L and (b) if $x_\infty \in S$ then L/S is compact. It is easily checked that L_∞ is a compact Hausdorff space inducing the original topology on L. Set

$$C_\infty^{\mathbb{R}}(L_\infty) = \{f \in C^{\mathbb{R}}(L_\infty) : f(x_\infty) = 0\}.$$

Then $C_\infty^{\mathbb{R}}(L_\infty)$ is a closed subalgebra of $C^{\mathbb{R}}(L_\infty)$. Furthermore, if L is a locally compact Hausdorff space then $f \mapsto f|L$ is an isometric isomorphism between $C_\infty^{\mathbb{R}}(L_\infty)$ and $C_0^{\mathbb{R}}(L)$. This is why the elements of $C_0^{\mathbb{R}}(L)$ are said to be the continuous functions on L *vanishing at* ∞.

These remarks imply the following reformulation of the Stone–Weierstrass theorem for real functions. We shall say that a set A of functions on a set S *separates the points of S strongly* if for all $x, y \in S$ $(x \neq y)$ there are functions, $f, g \in A$ such that $f(x) \neq f(y)$, $f(x) \neq 0$ and $g(y) \neq 0$. Note that by Theorem 9 if a subalgebra A of $C^{\mathbb{R}}(K)$ separates the points of K strongly then $\bar{A} = C^{\mathbb{R}}(K)$.

Theorem 9′. Let L be a locally compact Hausdorff space and let A be a subalgebra of $C_0^{\mathbb{R}}(L)$ strongly separating the points of L. Then A is dense in $C_0^{\mathbb{R}}(L)$, i.e. $\bar{A} = C_0^{\mathbb{R}}(L)$.   □

It is obvious that in Theorem 9′ we cannot replace $C_0^{\mathbb{R}}(L)$ by $C^{\mathbb{R}}(L)$ because $C_0^{\mathbb{R}}(L)$ itself is a uniformly closed subalgebra of $C^{\mathbb{R}}(L)$. Furthermore, we cannot replace $C_0^{\mathbb{R}}(L)$ by $C_0(L)$ either. This can be seen from the following example. Let $\Delta = \{z \in \mathbb{C} : |z| \leqslant 1\}$ be the closed unit disc in $\mathbb{C}$ and let A consist of those continuous functions on Δ that are analytic in the interior of Δ. Then A is a closed subalgebra of $C_0(\Delta) = C(\Delta)$, it strongly separates the points but does not contain the function $f(z) = \bar{z} \in C_0(\Delta)$.

The example above is, in fact, not an *ad hoc* example but one that goes to the heart of the matter: what we lack is complex conjugation. This leads us to a form of the Stone–Weierstrass theorem for complex functions, which is an easy consequence of the second variant, Theorem 9′.

Theorem 10. (Stone–Weierstrass theorem for complex functions) Let L be a locally compact Hausdorff space. Suppose A is a complex subalgebra of $C_0(L)$ which is closed under complex conjugation and strongly separates the points of L. Then $\bar{A} = C_0(L)$.

Proof. Let $A^{\mathbb{R}}$ be the set of real-valued functions in A. Then $A^{\mathbb{R}}$ is a real subalgebra of A and it is also a subalgebra of $C_0^{\mathbb{R}}(L)$. As $A^{\mathbb{R}}$ contains $(f+\bar{f})/2$ and $(f-\bar{f})/2i$ for every $f \in A$, it satisfies the conditions of Theorem 9'. Therefore, $C_0^{\mathbb{R}} \subset \bar{A}$ and so

$$iC_0^{\mathbb{R}}(L) \subset \bar{A} \quad \text{and} \quad C_0(L) = C_0^{\mathbb{R}}(L) + iC_0^{\mathbb{R}}(L) \subset \bar{A}. \qquad \square$$

Let us note some immediate consequences of the Stone–Weierstrass theorem; the first is the original theorem due to Weierstrass.

Corollary 11. Every continuous real-valued function on a bounded closed subset of $\mathbb{R}^n$ can be uniformly approximated by polynomials. $\square$

Corollary 12. Every continuous complex-valued function on the circle $T = \{z \in \mathbb{C} : |z| = 1\}$ can be uniformly approximated by trigonometric polynomials $\sum_{k=-n}^{n} c_k z^k$. $\square$

Corollary 13. Let K and L be compact Hausdorff spaces and endow $K \times L$ with the product topology. Then every continuous function $f: K \times L \to \mathbb{C}$ can be uniformly approximated by functions of the form $\sum_{i=1}^{n} g_i(x) h_i(y)$, where $g_i \in C(K)$ and $h_i \in C(L)$ $(1 \leq i \leq n)$. $\square$

The Stone–Weierstrass theorem is very useful in showing that various subspaces of function spaces are dense. As an example, let us look at the real space $L_p[0,1]$, where $1 \leq p < \infty$. The subspace of $L_p[0,1]$ consisting of continuous functions is dense. Since every continuous real-valued function on $[0,1]$ can be uniformly approximated by polynomials, and the uniform norm dominates the L_p-norm, the space of polynomials is also dense in $L_p[0,1]$.

Exercises

1. Show that a metric space is compact iff it is countably compact iff it is sequentially compact.
2. Show that a metric space is compact if and only if it is totally bounded and complete.
3. Show that a subset of a complete metric space is totally bounded if and only if its closure is compact.

4. A topological space X is *completely regular* if for every closed set $A \subset X$ and point b not in A there is a continuous function $f: X \to [0,1]$ such that $f|A = 0$ and $f(b) = 1$. Prove that the half-open interval topologies of the real line are completely regular.

5. Let A be a subset of a normal topological space X and let $f: A \to [0,1]$ be a continuous function. Does it follow that f has a continuous extension to the whole of X? And what is the answer if X is a compact Hausdorff space?

6. Let $C^{\mathbb{R}}(X)$ be the normed space of bounded continuous real-valued functions on a topological space X, with the uniform (supremum) norm. Show that $C^{\mathbb{R}}(X)$ is a Banach space.

7. Let V be a subspace of $C^{\mathbb{R}}(X)$ such that whenever A and B are non-empty disjoint closed subsets of X and $a, b \in \mathbb{R}$ then there is a function $f \in V$ such that $f(X) \subset [a,b]$, $f|A = a$ and $f|B = b$. Prove that V is dense in $C^{\mathbb{R}}(X)$.

8. Let K be a compact metric space and $S \subset C(K)$. Show that S is equicontinuous iff for every $\epsilon > 0$ there is a $\delta > 0$ such that $|f(x) - f(y)| < \epsilon$ whenever $d(x,y) < \delta$. Is this true for every metric space K?

9. Prove the following complex version of the Tietze–Urysohn extension theorem. Let A be a closed subset of a normal topological space X and let $f: A \to \mathbb{C}$ be a continuous function such that $\|f\|_A = \sup_{x \in A} |f(x)| = 1$. Show that there is a continuous function $F: X \to \mathbb{C}$ such that $\|F\|_X = \sup_{x \in X} |F(x)| = 1$.

10. Let L be a locally compact Hausdorff space. Show that a set $S \subset C_0(L)$ is totally bounded iff it is bounded, equicontinuous *and* for every $\epsilon > 0$ there is a compact set $K \subset L$ such that $|f(x)| < \epsilon$ for all $x \in L \backslash K$ and $f \in S$.

11. Let $B_r = \{z \in \mathbb{C} : |z| \leqslant r\}$, $D = \{z \in \mathbb{C} : |z| < 1\}$ and let F be a set of functions analytic on D such that $|f(z)| \leqslant M_r$ if $|z| = r < 1$. Show that for $r < 1$ the set $F|B_r = \{f|B_r : f \in F\}$ is a relatively compact subset of $C(B_r)$. Deduce that every sequence $(f_n)_1^\infty \subset F$ contains a subsequence $(f_{n_k})_1^\infty$ which is uniformly convergent on every B_r ($r < 1$) to a function f analytic on D.

12. Let U be an open subset of $\mathbb{C}$ and let $f_1, f_2, \ldots : U \to \mathbb{C}$ be uniformly bounded analytic functions. Prove that there is a subsequence $(f_{n_k})_1^\infty$ which is uniformly convergent on every compact subset of U.

13. Show that all three conditions are needed in Dini's theorem, i.e. that the conclusion $\|f_n - f_0\| \to 0$ in Theorem 5 need not hold if

only two of the following three conditions hold: (i) K is compact; (ii) the functions are continuous; (iii) $(f_n(x))_1^\infty$ is monotonic.

14. Let G be an open subset of $\mathbb{R}^2$ and let $f: G \to \mathbb{R}$ be continuous. Use the Arzelà–Ascoli theorem to prove Peano's theorem that for each point $(x_0, y_0) \in G$, at least one solution of

$$y'(x) = f(x, y)$$

passes through (x_0, y_0). [HINT: Let V be a closed neighbourhood of (x_0, y_0) and set $K = \sup\{|f(x, y)| : (x, y) \in V\}$. Choose $a < x_0 < b$ such that

$$\left\{ (x, y) : x \in [a, b], x \neq x_0, \left| \frac{y - y_0}{x - x_0} \right| < K \right\} \subset U.$$

The aim is to show that our differential equation has a solution through (x_0, y_0), defined on $[a, b]$. Find such a solution by considering piecewise linear approximations, say with division points $x_0 \pm k/n$ $(k = 1, 2, \ldots)$.]

15. Let K be a compact metric space. Prove that $C(K)$ is separable.

16. Let K be a compact Hausdorff space. Suppose $C(K) = \bigcup_{n=1}^\infty C_n$, where each C_n is an equicontinuous set of functions. What can you say about K?

17[+]. We say that $\epsilon > 0$ is a *Lebesgue number* of a cover $\bigcup_{\gamma \in \Gamma} U_\gamma$ of a metric space if, for every point x, the ball $B(x, \epsilon)$ is contained in some U_γ. Show that a metric space (X, d) is compact iff for every metric equivalent to d every open cover has a (positive) Lebesgue number.

18. Let K be a compact metric space, L a metric space, and let $C(K,L)$ to be set of continuous mappings from K to L with the uniform metric

$$d(f, g) = \sup_{x \in K} d(f(x), g(x)) = \max_{x \in K} d(f(x), g(x)).$$

Show that a subset S of $C(K,L)$ is totally bounded if and only if it is equicontinuous (i.e. for all $x \in K$ and $\epsilon > 0$ there is a $\delta > 0$ such that if $d(x, y) < \delta$ then $d(f(x), f(y)) < \epsilon$ for every $f \in S$) and the set $\{f(x) : f \in S\}$ is a totally bounded subset of L for every $x \in K$.

19[+]. Construct a countable Hausdorff space on which every continuous function is constant. The first such example was given by Urysohn in 1925.

20. Let K be a compact Haudsorff space. What are the maximal ideals of $C^{\mathbb{R}}(K)$? And the maximal closed subalgebras?

21. Let K be a compact Hausdorff space. For a continuous surjection φ of K onto a compact Hausdorff space K', let $\varphi^* : C^{\mathbf{R}}(K') \to C^{\mathbf{R}}(K)$ be given by $\varphi^*(f)(x) = (f\varphi)(x) = f(\varphi(x))$. Show that $\operatorname{Im}\varphi^*$ is a closed unital subalgebra of $C^{\mathbf{R}}(K)$ and every closed unital subalgebra of $C^{\mathbf{R}}(K)$ can be obtained in this way. Give a similar description of the closed subalgebras of $C^{\mathbf{R}}(K)$.

22. Let X be a normal topological space and let $U_1, \ldots, U_n$ be open sets covering X. Prove that there are continuous functions $f_1, \ldots, f_n : X \to [0,1]$ such that $\sum_{k=1}^{n} f_k(x) = 1$ for every x and $f_k(x) = 0$ for $x \notin U_k$ $(k = 1, \ldots, n)$. (The functions $f_1, \ldots, f_n$ are said to form a *partition of unity subordinate to the cover* $\{U_1, \ldots, U_n\}$.)

Notes

The relevant results of Urysohn and Tietze are in P. Urysohn, *Über die Mächtigkeit der zusammenhängenden Mengen*, Math. Ann., **94** (1925), 262–95 and H. Tietze, *Über Funktionen, die auf einer abgeschlossenen Menge stetig sind*, J. für die reine und angewandte Mathematik, **145** (1915), 9–14. Heinrich Tietze proved a special case of the extension theorem, while Paul Urysohn (who was Russian but tended to publish in German) proved Theorems 2 and 3 in the generality we stated them. Various special cases of the extension theorem were proved by a good many people, including Lebesgue, Brouwer, de la Vallée Poussin, Bohr and Hausdorff. For the Arzelà–Ascoli theorem see C. Arzelà, *Sulle serie di funzioni (parte prima)*, Memorie Accad. Sci. Bologna, **8** (1900), 131–86. Of course, there are several excellent books on general topology; J. L. Kelley, *General Topology*, Van Nostrand, Princeton, New Jersey, 1963, xiv and 298 pp., is particularly recommended.

The original version of Theorem 9 is in K. T. Weierstrass, *Über die analytische Darstellbarkeit sogenannter willkürlicher Funktionen reeler Argumente*, S.-B. Deutsch Akad. Wiss. Berlin, Kl. Math. Phys. Tech., 1885, 633–39 and 789–805. The Stone–Weierstrass theorem itself is in M. H. Stone, *Applications of the theory of Boolean rings to general topology*, Trans. Amer. Math. Soc., **41** (1937), 375–81; an exposition of the field can be found in M. H. Stone, *The generalized Weierstrass approximation theorem*, Math. Mag., **21** (1948), 167–84.

7. THE CONTRACTION-MAPPING THEOREM

This chapter is more or less an interlude in our study of normed spaces. Given a set S and a function $f: S \to S$, a point $x \in S$ satisfying $f(x) = x$ is said to be a *fixed point* of f. There is a large and very important body of 'fixed-point theorems' in analysis, that is, results claiming that every function satisfying certain conditions has a fixed point. Theorems of this kind often enable us to solve equations satisfying rather weak conditions. The aim of this section is to present the most fundamental of these fixed-point theorems and some of its great many applications. In Chapter 15 we shall return to the topic to prove some more sophisticated results.

A map $f: X \to X$ of a metric space into itself is said to be a *contraction* if

$$d(f(x), f(y)) \leqslant kd(x, y) \qquad (1)$$

for some $k < 1$ and all $x, y \in X$. One also calls f a *contraction with constant* k. It is immediate from the definition that every contraction map is continuous; in fact, it is uniformly continuous. Although we shall not make much use of this terminology, a function f between metric spaces is said to satisfy the *Lipschitz condition with constant* k if (1) holds for all x and y. Thus a map f from a metric space into itself is a contraction map if it satisfies a Lipschitz condition with constant less than 1.

The result below is usually referred to as *Banach's fixed-point theorem* or the *contraction-mapping theorem*.

Theorem 1. Let $f: X \to X$ be a contraction of a (non-empty) complete metric space. Then f has a unique fixed point.

Proof. Suppose that $k < 1$ and f satisfies (1) for all $x, y \in X$. Pick a point $x_0 \in X$ and set $x_1 = f(x_0)$, $x_2 = f(x_1)$, and so on: for $n \geqslant 1$ set $x_n = f(x_{n-1})$. Writing d for $d(x_0, x_1)$, we find that

$$d(x_n, x_{n+1}) \leqslant k^n d. \tag{2}$$

Indeed, (2) is trivially true for $n = 1$; assuming that it holds for n, we have

$$d(x_{n+1}, x_{n+2}) = d(f(x_n), f(x_{n+1})) \leqslant k d(x_n, x_{n+1}) \leqslant k^{n+1} d.$$

The triangle inequality and (2) imply that if $n < m$ then

$$d(x_n, x_m) \leqslant d(x_n, x_{n+1}) + d(x_{n+1}, x_{n+2}) + \ldots + d(x_{m-1}, x_m)$$
$$\leqslant \sum_{i=0}^{m-n-1} k^{n+i} d$$
$$\leqslant \frac{k^n}{1-k} d.$$

Hence $(x_n)_0^\infty$ is a Cauchy sequence and so it converges to some point $x \in X$. Since f is continuous, the sequence $(f(x_n))_0^\infty$ converges to $f(x)$. But $(f(x_n))_0^\infty = (x_n)_1^\infty$ converges to x and so $f(x) = x$. The uniqueness of the fixed point is even simpler: if $f(x) = x$ and $f(y) = y$ then $d(x, y) = d(f(x), f(y)) \leqslant k d(x, y)$, and so $d(x, y) = 0$, i.e. $x = y$. $\square$

When looking for a fixed point of a contraction map f, it is often useful to remember that the fixed point is the limit of *every* sequence $(x_n)_0^\infty$ consisting of the iterates of a point x_0, i.e. defined by picking a point x_0 and setting $x_n = f(x_{n-1})$ for $n \geqslant 1$. Let us note the following slight extension of Theorem 1.

Theorem 2. Let X be a complete metric space and let $f: X \to X$ be such that f^n is a contraction map for some $n \geqslant 1$. Then f has a unique fixed point.

Proof. We know from Theorem 1 that f^n has a unique fixed point x, say. Since

$$f^n(f(x)) = f(f^n(x)) = f(x),$$

$f(x)$ is also a fixed point of f and so $f(x) = x$. As a fixed point of f is also a fixed point of f^n, the map f does have a unique fixed point. $\square$

The very simple results above and their proofs have a large number of applications. In the rest of this chapter we shall be concerned with two directions: first we shall study maps between Banach spaces and then we shall present some applications to integral equations.

The proof of Theorem 1, sometimes called the *method of successive approximations*, is very important in the study of maps between Banach spaces. Here is a standard example concerning functions of two variables which are Lipschitz in their second variable. Recall that for a normed space X we write $D_r(X) = \{x \in X : \|x\| < r\}$ for the open ball of centre 0 and radius r.

Theorem 3. Let X and Y be Banach spaces and let

$$F: D_r(X) \times D_s(Y) \to Y$$

be a continuous map such that

$$\|F(x, y_1) - F(x, y_2)\| \leqslant k\|y_1 - y_2\| \tag{3}$$

and

$$\|F(x, 0)\| < s(1 - k) \tag{4}$$

for all $x \in D_r(X)$ and $y_1, y_2 \in D_s(Y)$, where $r, s > 0$ and $0 < k < 1$. Then there is a unique map $y : D_r(X) \to D_s(Y)$ such that

$$y(x) = F(x, y(x)) \tag{5}$$

for every $x \in D_r(X)$, and this map y is continuous.

Proof. For $x \in D_r(X)$, define $y_0(x) = 0$, $y_1(x) = F(x, y_0(x)) = F(x, 0)$ and $d = \|y_1(x) - y_0(x)\| = \|y_1(x)\|$. Note that, by (4), $d < s(1-k)$. Furthermore, if $n \geqslant 1$ and $y_n(x) \in D_s(Y)$ then set

$$y_{n+1}(x) = F(x, y_n(x)).$$

We claim that $y_n(x)$ is defined for every n, and that

$$\|y_{n+1}(x) - y_n(x)\| \leqslant k^n d < k^n s(1 - k) \tag{6}$$

and

$$\|y_{n+1}(x)\| \leqslant \sum_{i=0}^{n} k^i d \leqslant \frac{d}{1-k} < s. \tag{7}$$

As in the proof of Theorem 1, this follows by induction on n. Indeed, (6) clearly holds for $n = 0$; assuming that $y_0(x), y_1(x), \ldots, y_n(x) \in D_s(Y)$ and (6) holds for $n - 1$, inequality (3) implies that

$$\|y_{n+1}(x) - y_n(x)\| = \|F(x, y_n(x)) - F(x, y_{n-1}(x))\|$$

$$\leqslant k\|y_n(x) - y_{n-1}(x)\| \leqslant k^n d.$$

In turn, this implies that (7) holds for n:

$$\|y_{n+1}(x)\| \leq \|y_0(x)\| + \sum_{i=0}^{n} \|y_{i+1}(x) - y_i(x)\|$$

$$\leq \sum_{i=0}^{n} k^i d$$

$$\leq \frac{d}{1-k} < s,$$

and so, in particular, $y_{n+1}(x) \in D_s(X)$.

As F is a continuous function, each $y_n : D_r(X) \to D_s(Y)$ is continuous. Inequalities (6) and (7) imply that the sequence $(y_n)_0^\infty$ is uniformly convergent to a continuous function y from $D_r(X)$ to $D_s(Y)$.

Then (5) holds trivially:

$$y(x) = \lim_{n \to \infty} y_n(x) = \lim_{n \to \infty} F(x, y_{n-1}(x)) = F(x, y(x)).$$

Furthermore, $y(x)$ is unique, since if $y_0(x)$ is a solution of (5) then

$$\|y(x) - y_0(x)\| = \|F(x, y(x)) - F(x, y_0(x))\| \leq k\|y(x) - y_0(x)\|,$$

and so $y(x) = y_0(x)$. $\qquad\qquad\qquad\qquad\qquad\qquad\qquad\qquad\qquad\square$

The next result claims that a suitable map close to the identity map is, in fact, a homeomorphism.

Theorem 4. Let X be a Banach space and let $\epsilon : D_s(X) \to X$ be a contraction with constant $k < 1$ such that

$$\|\epsilon(0)\| < r,$$

where $s > 0$ and $r = \frac{1}{2}s(1-k)$. Define $f : D_s(X) \to X$ by

$$f(x) = x + \epsilon(x).$$

Then there is an open neighbourhood U of 0 such that the restriction of f to U is a homeomorphism of U onto $D_r(X)$.

Proof. Define $F : D_r(X) \times D_s(X) \to X$ by

$$F(x, y) = x - \epsilon(y).$$

Then F satisfies the conditions of Theorem 3. Indeed, (3) holds by assumption; furthermore, if $x \in D_r(X)$ then

$$\|F(x, 0)\| = \|x - \epsilon(0)\| < r + r = s(1 - k),$$

and so (4) also holds. Hence, by Theorem 3, there is a unique function

$g : D_r(X) \to D_s(X)$ such that

$$g(x) = F(x, g(x)) = x - \epsilon(g(x)),$$

i.e.

$$x = g(x) + \epsilon(g(x)) = f(g(x))$$

for all $x \in D_r(X)$. Set $U = g(D_r(X))$. To complete the proof, we have to check that U is an open neighbourhood of 0 and $f : U \to D_r(X)$ is a homeomorphism. Clearly, f is a one-to-one map of U onto $D_r(X)$. Since g is unique, the continuity of f implies that $U = f^{-1}(D_r(X))$ is open. As g is also continuous, f is indeed a homeomorphism. Finally, $f(0) = \epsilon(0) \in D_r(X)$ and so $0 \in M U$. □

From Theorem 4 it is a small step to the inverse-function theorem for Banach spaces.

In defining differentiable maps, it is convenient to use the 'little oh' notation. Let X and Y be Banach spaces. Given a function $\alpha : D \to Y$, where $D \subset X$, we write

$$\alpha(h) = o(h)$$

if for every $\epsilon > 0$ there is a $\delta > 0$ such that

$$\|\alpha(h)\| \leqslant \epsilon \|h\|$$

whenever $h \in D$ and $\|h\| < \delta$. Thus '$\alpha(h)$ tends to 0 faster than h'.

Let U be an open subset of X and let $f : U \to Y$. We call f *differentiable at* $x_0 \in U$ if there is a linear operator $T \in \mathcal{B}(X, Y)$ such that

$$f(x_0 + h) = f(x_0) + Th + o(h).$$

The operator T is the *derivative* of f at x_0: in notation, $f'(x_0) = T$. Thus f is differentiable at x_0 with derivative $T \in \mathcal{B}(X, Y)$ iff for every $\epsilon > 0$ there is a $\delta > 0$ such that

$$\|f(x_0 + h) - f(x_0) - Th\| \leqslant \epsilon \|h\|$$

whenever $\|h\| < \delta$.

Note that a linear operator $T \in \mathcal{B}(X, Y)$, $x \mapsto Tx$, is differentiable at every point, and $T'(x) = T$ for every $x \in X$. We shall need the following simple analogue of the mean-value theorem.

Lemma 5. Let X and Y be Banach spaces, $U \subset X$ an open convex set and $f : U \to Y$ a differentiable function, with

$$\|f'(x)\| \leqslant k$$

for all $x \in U$. Then

$$\|f(x) - f(y)\| \leqslant k\|x - y\|$$

for all $x, y \in U$.

Proof. If the assertion is false then one can find two sequences $(x_n)_0^\infty, (y_n)_0^\infty \subset U$ such that $x_0 \neq y_0$, and for $n \geqslant 0$ either $x_{n+1} = x_n$ and $y_{n+1} = \frac{1}{2}(x_n + y_n)$, or $x_{n+1} = \frac{1}{2}(x_n + y_n)$ and $y_{n+1} = y_n$, and

$$\|f(x_n) - f(y_n)\| \geqslant k'\|x_n - y_n\|$$

for some $k' > k$ and all $n \geqslant 0$. But then there is a $z_0 \in U$ such that $x_n \to z_0$, $y_n \to z_0$ and

$$\|x_n - y_n\| = \|x_n - z_0\| + \|z_0 - y_n\|,$$

as the segments $[x_n, y_n]$ are nested and $\|x_{n+1} - y_{n+1}\| = \frac{1}{2}\|x_n - y_n\|$. Hence

$$
\begin{aligned}
k'\|x_n - y_n\| &\leqslant \|f(x_n) - f(y_n)\| \\
&\leqslant \|f(x_n) - f(z_0)\| + \|f(z_0) - f(y_n)\| \\
&\leqslant \|f'(z_0)(x_n - z_0)\| + o(x_n - z_0) \\
&\qquad + \|f'(z_0)(z_0 - y_n)\| + o(z_0 - y_n) \\
&\leqslant k\|x_n - z_0\| + k\|z_0 - y_n\| + o(x_n - y_n) \\
&= k\|x_n - y_n\| + o(x_n - y_n),
\end{aligned}
$$

which is a contradiction, as $\|x_n - y_n\| \neq 0$ and $\|x_n - y_n\| \to 0$.   ☐

Like Theorem 4, the following theorem ensures that, under suitable circumstances, a function f is a homeomorphism in some neighbourhood.

Theorem 6. Let X and Y be Banach spaces, let U be an open neighbourhood of a point $x_0 \in X$ and let $f: U \to Y$ be such that $f'(x)$ exists and is continuous in U. Suppose

$$(f'(x_0))^{-1} \in \mathcal{B}(Y, X).$$

Then f is a homeomorphism of an open neighbourhood U_0 of x_0 onto an open neighbourhood V_0 of $f(x_0)$, f^{-1} is continuously differentiable on V_0 and

$$(f^{-1})'(y) = (f'(f^{-1}(y)))^{-1}$$

for $y \in V_0$.

Proof. We may assume that $x_0 = 0$ and $f(x_0) = 0$. Furthermore, setting $T_0 = f'(0)$ and replacing f by $T_0^{-1} \circ f : X \to X$, we may assume that $Y = X$ and $f'(0)$ is the identity operator I.

Set $\epsilon(x) = f(x) - x$. Then $\epsilon : U \to Y$ is continuously differentiable and $\epsilon'(0) = 0$. Hence there is an open ball $D_s(X) \subset U$ such that

$$\|\epsilon'(x)\| \leq \tfrac{1}{2}$$

for $x \in D_s(X)$. Then, by Lemma 5,

$$\|\epsilon(x) - \epsilon(y)\| \leq \tfrac{1}{2}\|x - y\|$$

for all $x, y \in D_s(X)$. Hence, by Theorem 4, $f(x) = x + \epsilon(x)$ is a homeomorphism of some open neighbourhood U_0 of 0 onto $V_0 = D_{s/4}(X)$.

The rest is plain sailing. For $y_0 \in V_0$ set $x_0 = f^{-1}(y_0) \in U_0$ and $T = f'(x_0)$. Furthermore, for $y_0 + k \in V_0$ set $x_0 + h = f^{-1}(y_0 + k)$. As $k \to 0$, we have $h \to 0$, and so $k = Th + o(h)$ and

$$h = T^{-1}k + o(k).$$

Consequently,

$$f^{-1}(y_0 + k) = f^{-1}(f(x_0 + h)) = x_0 + h = x_0 + T^{-1}k + o(k),$$

proving that $(f^{-1})'(y_0) = T^{-1} = (f'(f^{-1}(y_0)))^{-1}$. The continuity of $(f^{-1})'$ follows from the fact that the map $T \mapsto T^{-1}$ is continuous. $\square$

The contraction-mapping theorem has numerous applications to differential and integral equations. Here we shall confine ourselves to three rather simple examples.

Theorem 7. Let $K(s, t)$ be a continuous real function on the unit square $[0, 1]^2$, and let $v(s)$ be a continuous real function on $[0, 1]$. Then there is a unique continuous real function $y(s)$ on $[0, 1]$ such that

$$y(s) = v(s) + \int_0^s K(s, t) y(t) \, dt.$$

Proof. We shall consider various operators on the Banach space $C[0, 1]$ with the uniform (supremum) norm. The linear integral operator $L : C[0, 1] \to C[0, 1]$, defined by

$$(Lx)(s) = \int_0^s K(s, t) x(t) \, dt$$

is called the *Volterra integral operator with kernel* $K(s,t)$. In fact, we could assume that $K(s,t)$ is defined only on the closed triangle $0 \leqslant t \leqslant s \leqslant 1$, since for $s < t$ we do not use $K(s,t)$. Write $K_1(s,t) = K(s,t)$ and for $n \geqslant 1$ set

$$K_{n+1}(s,t) = \int_t^s K(s,u)K_n(u,t)du.$$

Here and in the rest of the proof we take an integral

$$\int_t^s f(u)\, du$$

to be 0 if $t > s$. Alternatively, we may take each $K_n(s,t)$ to be 0 outside the closed triangle $0 \leqslant t \leqslant s \leqslant 1$. It is easily shown by induction on n that L^n is precisely the Volterra integral operator with kernel K_n. Indeed, if this is true for n then

$$(L^{n+1}x)\,(s) = \int_0^s K(s,u) \int_0^u K_n(u,t)x(t)\, dt\, du$$

$$= \int_0^s \int_0^u K(s,u)\, K_n(u,t)x(t)\, dt\, du$$

$$= \int_0^u \int_t^s K(s,u)\, K_n(u,t)\, du\, x(t)\, dt$$

$$= \int_0^u K_{n+1}(s,t)x(t)\, dt.$$

The function $K(s,t)$ is continuous on the closed unit square and so it is bounded, say

$$|K(s,t)| \leqslant M$$

for all $0 \leqslant s,t \leqslant 1$. Then, again by induction on n, we see that

$$|K_n(s,t)| \leqslant \frac{M^n|s-t|^{n-1}}{(n-1)!}$$

for all $n \geqslant 1$. Indeed, if this holds for n then

$$|K_{n+1}(s,t)| \leqslant \int_t^s \frac{MM^n|u-t|^{n-1}}{(n-1)!}\, du$$

$$\leqslant \frac{M^{n+1}|s-t|^n}{n!}.$$

Hence if n is sufficiently large, say $n \geq 4M$, then

$$|K_n(s,t)| \leq 1$$

for all $0 \leq s, t \leq 1$. Therefore

$$\|(L^n x)(s)\| = \int_0^s K_n(s,t)|x(t)|dt \leq \|x\| \int_0^s |K_n(s,t)| \; dt \leq \tfrac{1}{2}\|x\|$$

and so

$$\|L^n\| \leq \tfrac{1}{2}.$$

In particular, L^n is a contraction.

It is time to return to our equation. Define a continuous (affine) operator $T: C[0,1] \rightarrow C[0,1]$ by $Tx = v + Lx$. The theorem claims that T has precisely one fixed point. But

$$T^n x = \left(\sum_{i=0}^{n-1} L^i\right)v + L^n x$$

and so T^n is a contraction. Hence, by Theorem 2, T has a unique fixed point. $\qquad\qquad\square$

The next application concerns a non-linear variant of the Volterra integral operator.

Theorem 8. Let $K(s,t)$ and $w(s,t)$ be continuous real functions on the unit square $[0,1]^2$, and let $v(s)$ be a continuous real function on $[0,1]$. Suppose that

$$|w(s,t_1) - w(s,t_2)| \leq N|t_1 - t_2|$$

for all $0 \leq t_1, t_2, s \leq 1$. Then there is a unique continuous real function $y(s)$ on $[0,1]$ such that

$$y(s) = v(s) + \int_0^s K(s,t)\,w(t,y(t))\,dt.$$

Proof. Define $L: C[0,1] \rightarrow C[0,1]$ by

$$(Lx)(s) = \int_0^s K(s,t)\,w(t,x(t))\,dt.$$

The theorem claims that the map $T: C[0,1] \rightarrow C[0,1]$, defined by

$$Tx = v + Lx,$$

has a unique fixed point.

To show this, we shall turn to trickery. For $a > 0$, we introduce a new norm $\|\cdot\|_a$ on $C[0,1]$:

$$\|x\|_a = \int_0^1 e^{-as}|x(s)|\,ds.$$

Then $\|\cdot\|_a$ is indeed a norm on $C[0,1]$, and it is equivalent to the L_1 norm. Set $X_a = (C[0,1], \|\cdot\|_a)$ and let $\bar{X}_a$ be the completion of X_a. Clearly, $\bar{X}_a$ is the vector space $L_1(0,1)$ with the norm $\|\cdot\|_a$, and L extends to a map $\bar{L}: \bar{X}_a \to \bar{X}_a$ given by the formula for L. Furthermore, with $M = \max\{|K(s,t)|: 0 \leqslant s,t \leqslant 1\}$ we have, for $x,y \in X_a = (L_1(0,1), \|\cdot\|_a)$,

$$\|\bar{L}x - \bar{L}y\|_a = \int_0^1 e^{-as}\left|\int_0^s K(s,t)\{w(t,x(t)) - w(t,y(t))\}\,dt\right|\,ds$$

$$\leqslant MN \int_0^1 \int_0^s e^{-as}|x(t)-y(t)|\,dt\,ds$$

$$= MN \int_0^1 \int_t^1 e^{-as}|x(t)-y(t)|\,ds\,dt$$

$$= MN \int_0^1 \frac{e^{-at}-e^{-a}}{a}|x(t)-y(t)|\,dt$$

$$\leqslant \frac{MN}{a}\|x-y\|_a.$$

This shows that for $a > MN$ the map

$$\bar{L}: \bar{X}_a \to \bar{X}_a$$

is a contraction, and so is $\bar{T} = v + T$. It is easily checked that $\bar{T}$ maps $\bar{X}_a$ into $X_a = (C[0,1], \|\cdot\|_a)$, so its unique fixed point belongs to $C[0,1]$, and is also the unique fixed point of T. $\qquad\square$

In our final example we have to do even less work, since the kernel satisfies a Lipschitz-type condition, just like the function $w(s,t)$ in Theorem 8.

Theorem 9. Let $K(s,t,u)$ be a continuous function on $0 \leqslant s,t \leqslant 1$, $u \geqslant 0$, such that

$$|K(x,t,u_1) - K(s,t,u_2)| \leqslant N(s,t)|u_1-u_2|,$$

where $N(s,t)$ is a continuous function satisfying

$$\int_0^1 N(s,t)\,dt \leqslant k < 1,$$

for every $0 \leqslant s \leqslant 1$. Then for every $v \in C[0,1]$ there is a unique function $y \in C[0,1]$ such that

$$y(s) = v(s) + \int_0^1 K(s,t,y(t)) \, dt.$$

Proof. Define $L: C[0,1] \to C[0,1]$ by

$$(Lx)(s) = \int_0^1 K(s,t,x(t)) \, dt,$$

so that a solution of our integral equation is precisely a fixed point of the map $T: C[0,1] \to C[0,1]$, given by $Tx = v + Lx$. For $x, y \in C[0,1]$ we have

$$|(Lx - Ly)(s)| = \left| \int_0^1 \{K(s,t,x(t)) - K(s,t,y(t))\} \, dt \right|$$

$$\leqslant \int_0^1 N(s,t) |x(t) - y(t)| \, dt$$

$$\leqslant k \|x - y\|.$$

Hence $\|Lx - Ly\| \leqslant k \|x - y\|$, implying that L is a contraction and therefore so is T. Consequently, by Theorem 1, T has a unique fixed point. $\square$

As we remarked earlier, in chapter 15 we shall prove several other fixed-point theorems. But for the time being we return to the study of normed spaces.

Exercises

1. Show that a contraction of an incomplete metric space into itself need not have a fixed point.
2. Let A and B be disjoint subsets of a metric space, with A compact and B closed. Show that

   $$d(A,B) = \inf\{d(x,y) : x \in A, y \in B\} > 0.$$

   Show also that this need not be true if we assume only that A and B are closed.
3. Show that if f is a map of a complete metric space X into itself such that $d(f(x), f(y)) < d(x,y)$ for all $x, y \in X$ $(x \neq y)$ then f

need not have a fixed point. Show however that if X is compact then f has a unique fixed point.

4. Let $f: X \to X$ be a map from a complete metric space into itself. Suppose that for every $\epsilon > 0$ there is a $\delta > \epsilon$ such that if $\epsilon \leqslant d(x,y) < \delta$ then $d(f(x),f(y)) < \epsilon$. Prove that f has a unique fixed point.

5. Let f be a map of a compact metric space X into itself. Suppose that for every $\epsilon > 0$ there is a $\delta = \delta(\epsilon) > 0$ such that if $d(x,f(x)) < \delta$ then $f(B(x,\epsilon)) \subset B(x,\epsilon)$. Let $x_0 \in X$ and define $(x_n)_0^\infty$ as in the proof of Theorem 1: $x_n = f(x_{n-1})$ for $n \geqslant 1$. Show that if $d(x_n, x_{n+1}) \to 0$ as $n \to \infty$ then the sequence $(x_n)_0^\infty$ converges to a fixed point of f.

6. Let f be a map of a complete metric space X into itself such that

$$d(f(x), f(y)) \leqslant \varphi(d(x,y))$$

for all $x, y \in X$, where $\varphi : \mathbb{R}^+ \to \mathbb{R}^+$ is a monotone increasing function such that $\lim_{n \to \infty} \varphi^n(t) = 0$ for every $t > 0$. Deduce from the result in the previous exercise that f has a unique fixed point a, and $\lim_{n \to \infty} f^n(x) = a$ for every $x \in X$.

7. Let f be a continuous map of a (non-empty) compact Hausdorff space into itself. Show that there is a (non-empty) compact subset such that $f(A) = A$.

8. Show that on the space c_0 the norm function $x = (x_i)_1^\infty \mapsto \|x\| = \sup|x_i|$ is differentiable at x iff $|x_i|$ has a unique maximum (i.e. there is an index j such that $|x_j| > |x_i|$ if $i \neq j$).

9. Show that on the space l_1 the norm function

$$x = (x_i)_1^\infty \mapsto \|x\| = \sum_{i=1}^\infty |x_i|$$

is not differentiable at any point.

In the next five exercises, X, Y and Z are Banach spaces.

10. Let $f, g : X \to Y$ be differentiable at $x_0 \in X$ and let α and β be constants. Show that $\alpha f + \beta g$ is differentiable at x_0 and

$$(\alpha f + \beta g)'(x_0) = \alpha f'(x_0) + \beta g'(x_0).$$

11. Let $f: U \to V$ be a homeomorphism of an open set $U \subset X$ onto an open set $V \subset Y$. Suppose that f is differentiable at $x_0 \in U$ and $(f'(x_0))^{-1} \in \mathscr{B}(Y,X)$. Show that $f^{-1}: V \to U$ is differentiable at

$y_0 = f(x_0)$ and

$$(f^{-1})'(y_0) = (f'(x_0))^{-1}.$$

12. Let $U \subset X$ and $V \subset Y$ be open sets, and $f: U \to Y$ and $g: V \to Z$ continuous maps with $f(U) \subset V$. Suppose that f is differentiable at a point $x_0 \in U$ and g is differentiable at $y_0 = f(x_0) \in V$. Show that the map $g \circ f: U \to Z$ is differentiable at x_0 and

$$(g \circ f)'(x_0) = g'(y_0) \circ f'(x_0).$$

13. Let U be a connected open subset of X and, for $n = 1, 2, \ldots$, let $f_n: U \to Y$ be differentiable at every point of U. Suppose that every $x_0 \in U$ has a neighbourhood $N(x_0)$ on which $\sum_{n=1}^{\infty} \|f_n'(x)\|$ is uniformly convergent. Show that if $\sum_{n=1}^{\infty} f_n(x)$ is convergent for some $x \in U$ then it is convergent for all $x \in U$, the map

$$f(x) = \sum_{n=1}^{\infty} f_n(x)$$

is differentiable at every $x \in U$ and

$$f'(x) = \sum_{n=1}^{\infty} f_n'(x).$$

14. Let $U \subset X$ be an open convex set and $f: U \to Y$ differentiable. Show that for all $x, y, z \in U$ we have

$$\|f(x) - f(y) - f'(z)(x - y)\| \leq \|x - y\| \sup_{s \in [x, y]} \|f'(s) - f'(z)\|,$$

where, as usual, $[x, y]$ is the *closed segment* $\{tx + (1 - t)y : 0 \leq t \leq 1\}$.

Notes

The contraction-mapping theorem is from S. Banach, *Sur les opérations dans les ensembles abstraits et leur applications aux équations intégrales*, Fundamenta Mathematica, **3** (1922), 133–81. Of course, it can be found in most books on general topology and basic functional analysis. A rich source of applications to differential and integral equations is D. H. Griffel, *Applied Functional Analysis*, Ellis Harwood, Chichester, 1985, 390 pp.

8. WEAK TOPOLOGIES AND DUALITY

When studying a normed space or various sets of linear operators on a space, it is often useful to consider topologies other than the norm topology we have introduced so far. Two of the most important of these topologies are the weak topology and the weak-star topology. Before we define these topologies and prove Alaoglu's theorem, the fundamental result concerning the weak-star topology, we shall review some additional concepts in point set topology and present a proof of Tychonov's theorem.

Given a topological space (X, τ), a collection σ of subsets of X is said to be a *sub-basis* for τ if $\sigma \subset \tau$ and every member of τ is a union of finite intersections of sets from σ. Equivalently, σ is a sub-basis for τ if $\sigma \subset \tau$ and for all $U \in \tau$ and $x \in U$ there are sets $S_1, \dots, S_n$ such that $x \in \bigcap_{i=1}^{n} S_i \subset U$. More formally, σ is a sub-basis for τ if

$$\tau = \left\{ \bigcup_{F \in \mathcal{F}} \bigcap_{S \in F} S : \mathcal{F} \subset \sigma^{(<\omega)} \right\} \cup \{\varnothing, X\}, \tag{1}$$

where $\sigma^{(<\omega)}$ is the set of all finite subsets of σ. In most cases we may omit the second term on the right, $\{\varnothing, X\}$, especially if the intersection of an empty collection of subsets of X is taken to be X. Not only do sub-bases tend to be more economical than bases, but they also enable us to use an arbitrary family of sets to define a topology. Indeed, if $\sigma \subset \mathscr{P}(X)$ is any collection of subsets of X then the collection τ defined by (1) is a topology on X.

The possibility of using an arbitrary set system to define a topology enables us to use a set of functions into topological spaces to define a topology on a set. Let X be a set and for each $\gamma \in \Gamma$ let f_γ be a map of X into a topological space (X_γ, τ_γ). Then there is a unique weakest topology τ on X such that every $f_\gamma : (X, \tau) \to (X_\gamma, \tau_\gamma)$ is continuous. Indeed, taking

$$\sigma = \{f_\gamma^{-1}(U_\gamma) : U_\gamma \subset X_\gamma \text{ is open in } \tau_\gamma \ (\gamma \in \Gamma)\},$$

the topology τ is given by (1). Thus $U \subset X$ is open in τ iff for every $x \in U$ there are $\gamma_1, \ldots, \gamma_n \in \Gamma$ and $U_1 \in \tau_{\gamma_1}, \ldots, U_n \in \tau_{\gamma_n}$, such that

$$x \in \bigcap_{i=1}^{n} f_{\gamma_i}^{-1}(U_{\gamma_i}) \subset U.$$

The topology τ defined in this way is called the *weak topology generated by* $\mathcal{F}$, and is denoted by $\sigma(X, \mathcal{F})$, where $\mathcal{F} = \{f_\gamma : \gamma \in \Gamma\}$.

A map f from a topological space (X', τ') into $(X, \sigma(X, \mathcal{F}))$ is continuous iff $f_\gamma \circ f : (X', \tau') \to (X_\gamma, \tau_\gamma)$ is continuous for every $\gamma \in \Gamma$. This so-called *universal property* characterizes the weak topology.

The product topology is one of the prime examples of a weak topology. Let (X_γ, τ_γ) be a topological space for each $\gamma \in \Gamma$ and let $X = \prod_{\gamma \in \Gamma} X_\gamma$ be the Cartesian product of the sets X_γ $(\gamma \in \Gamma)$. Thus X is the collection of all the functions $x : \Gamma \to \bigcup_{\gamma \in \Gamma} X_\gamma$ such that $x(\gamma) \in X_\gamma$. One usually writes x_γ for $x(\gamma)$ and sets $x = (x_\gamma)_{\gamma \in \Gamma}$; also, x_γ is the γ-*component* of x. Let $p_\gamma : X \to X_\gamma$ be the projection onto X_γ; thus $p_\gamma(x) = x_\gamma$. The *product topology* on X is the weak topology generated by the projections p_γ $(\gamma \in \Gamma)$. To spell it out: a set $U \subset X$ is open in the product topology iff for every point $x = (x_\gamma)_{\gamma \in \Gamma}$ there are indices $\gamma_1, \ldots, \gamma_n$ and open sets $U_{\gamma_1}, \ldots, U_{\gamma_n}$ $(U_{\gamma_i} \in \tau_{\gamma_i})$ such that $x_{\gamma_i} \in U_{\gamma_i}$ $(i = 1, \ldots, n)$ and

$$\bigcap_{i=1}^{n} p_{\gamma_i}^{-1}(U_{\gamma_i}) = \{y = (y_\gamma) \in X : y_{\gamma_i} \in U_{\gamma_i} \ (i = 1, \ldots, n)\} \subset U.$$

Thus to guarantee that a point $y = (y_\gamma)$ belongs to a fixed neighbourhood of $x = (x_\gamma)$, it suffices to demand that for finitely many indices $\gamma_1, \gamma_2, \ldots, \gamma_n$, each y_{γ_i} belongs to a certain neighbourhood of x_{γ_i}.

Let us turn to the main topic of this section, the weak and weak-star topologies of normed spaces. Given a normed space X with dual X^*, the *weak topology* $\sigma(X, X^*)$ is the weak topology on X generated by the bounded linear functionals on X. Thus $U \subset X$ is open in the weak topology (w-open) iff for every $x \in U$ there are bounded linear functionals $f_1, \ldots, f_n$ and positive reals $\epsilon_1, \ldots, \epsilon_n$, such that

$$\{y \in X : |f_i(x) - f_i(y)| < \epsilon_i \ (i = 1, \ldots, n)\} \subset U.$$

By replacing f_i by f_i / ϵ_i, we may take every ϵ_i to be 1.

The *weak-star topology* $\sigma(X^*, X)$ on the dual X^* of a normed space X is the weak topology generated by the elements $\hat{x} \in X^{**}$ for $x \in X$, i.e. by the elements of X acting on X^* as bounded linear functionals (see Theorem 3.10). In more detail: a set $G \subset X^*$ is open in the weak-star topology (w^*-open) iff for every $g \in G$ there are points $x_1, \ldots, x_n \in X$ and positive reals $\epsilon_1, \ldots, \epsilon_n$, such that

$$\{f \in X^* : |f(x_i) - g(x_i)| < \epsilon_i \ (i = 1, \ldots, n)\} \subset G.$$

As before, we may take every ϵ_i to be 1.

The Hahn–Banach theorem implies that if Y is a subspace of a normed space X then the restriction to Y of the weak topology on X is precisely the weak topology on Y, i.e.

$$\sigma(X, X^*)|Y = \sigma(Y, Y^*). \tag{2}$$

In view of the proof of Tychonov's theorem we shall give, it seems to be appropriate to mention Tukey's lemma, which is equivalent to Zorn's lemma (and so to the axiom of choice). A system $\mathcal{F}$ of subsets of a set S is said to be of *finite character* if a subset A of S belongs to $\mathcal{F}$ if and only if every finite subset of A belongs to $\mathcal{F}$.

Lemma 1. (Tukey's lemma) Let $\mathcal{F}$ be a set system of finite character and let $F \in \mathcal{F}$. Then $\mathcal{F}$ has a maximal element containing F.

Proof. Let $\mathcal{F}_0 = \{A \in \mathcal{F} : F \subset A\}$. Order the elements of $\mathcal{F}_0$ by inclusion: $A \leq B$ iff $A \subset B$. Then $\leq$ is a partial order on $\mathcal{F}_0$. Also, if $\mathcal{C} \subset \mathcal{F}_0$ is a chain in $\mathcal{F}_0$ then $D = \bigcup_{C \in \mathcal{C}} C$ belongs to $\mathcal{F}_0$ since all finite subsets of D belong to $\mathcal{F}$, and $F \subset D$. Clearly, $D \in \mathcal{F}_0$ is an upper bound for $\mathcal{C}$. Hence, by Zorn's lemma, $\mathcal{F}_0$ has a maximal element. $\square$

A system $\mathcal{F}$ of subsets of a set has the *finite-intersection property* if $\bigcap_{i=1}^n F_i \neq \varnothing$ for all $F_1, \ldots, F_n \in \mathcal{F}$. It is easily seen that a topological space X is compact iff the intersection of all subsets of a system $\mathcal{F}$ of closed sets which has the finite-intersection property is non-empty. Indeed, $\mathcal{G} = \{X \backslash F : F \in \mathcal{F}\}$ is a collection of open sets. Since no finite collection of sets in $\mathcal{G}$ covers X, the entire system $\mathcal{G}$ also fails to cover X, so there is a point $x \in X$ which belongs to no member of $\mathcal{G}$. Hence $x \in \bigcap_{F \in \mathcal{F}} F$.

A trivial reformulation of this characterization of compactness is the following; a topological space is compact iff whenever $\mathcal{A}$ is a system of subsets with the finite-intersection property then $\bigcap_{A \in \mathcal{A}} \bar{A} \neq \varnothing$. This is the characterization we shall use in the proof below.

Theorem 2. (Tychonov's theorem) The product of a collection of compact sets is compact.

Proof. Let $\{X_\gamma : \gamma \in \Gamma\}$ be a collection of compact spaces and let $X = \prod_{\gamma \in \Gamma} X_\gamma$ be endowed with the product topology. We have to show that X is compact, i.e. that if $\mathcal{A}$ is a system of subsets of X with the finite-intersection property then $\bigcap_{A \in \mathcal{A}} \bar{A} \neq \varnothing$.

Let $\mathcal{F}$ be the collection of all set systems on X with the finite-intersection property. Then $\mathcal{F}$ is of finite character and so, by Tukey's lemma, there is a maximal set system $\mathcal{B}$ with the finite-intersection property containing $\mathcal{A}$. Since

$$\bigcap_{B \in \mathcal{B}} \bar{B} \subset \bigcap_{A \in \mathcal{A}} \bar{A},$$

we may assume that our set system $\mathcal{A}$ is itself is maximal for the finite-intersection property.

The maximality of $\mathcal{A}$ implies that the intersection of any two sets in $\mathcal{A}$ belongs to $\mathcal{A}$, as does the intersection of finitely many sets in $\mathcal{A}$. Consequently, if $B \subset X$ meets every set in $\mathcal{A}$ then $B \in \mathcal{A}$. In particular, if $A \in \mathcal{A}$ and $A \subset B \subset X$ then $B \in \mathcal{A}$.

So far we have made no use of the fact that X is a product space, let alone a product of compact spaces. Let us do so now. For each $\gamma \in \Gamma$, the system $\{p_\gamma(A) : A \in \mathcal{A}\}$ of subsets of X_γ has the finite-intersection property. As X_γ is compact, for some point $x_\gamma \in X_\gamma$ we have

$$x_\gamma \in \bigcap_{A \in \mathcal{A}} \overline{p_\gamma(A)}.$$

This means precisely that for every neighbourhood U_γ of x_γ, the set $p_\gamma^{-1}(U_\gamma)$ meets every element of $\mathcal{A}$. Hence $p_\gamma^{-1}(U_\gamma) \in \mathcal{A}$ and so

$$\bigcap_{i=1}^{n} p_{\gamma_i}^{-1}(U_{\gamma_i}) \in \mathcal{A} \tag{3}$$

whenever $\{\gamma_1, \ldots, \gamma_n\} \subset \Gamma$ and U_{γ_i} is a neighbourhood of x_{γ_i}.

Let $x = (x_\gamma)_{\gamma \in \Gamma} \in X$. We claim that $x \in \bigcap_{A \in \mathcal{A}} \bar{A}$. Indeed, for every neighbourhood U of x there are indices $\gamma_1, \ldots, \gamma_n \in \Gamma$ and neighbourhoods U_{γ_i} of x_i ($i = 1, \ldots, n$) such that

$$\bigcap_{i=1}^{n} p_{\gamma_i}^{-1}(U_{\gamma_i}) \subset U.$$

Consequently, (3) implies that $U \in \mathcal{A}$, i.e. U intersects every element of $\mathcal{A}$. Hence every neighbourhood of x intersects every $A \in \mathcal{A}$, and so $x \in \bar{A}$ for every $A \in \mathcal{A}$. This proves our claim and completes the proof of our theorem. $\qquad\square$

From here it is a small step to Alaoglu's theorem, the fundamental result concerning the weak-star topology; in view of the fact that both the weak-star topology and the product topology are weak topologies, this is hardly surprising.

Theorem 3. (Alaoglu's theorem) The unit ball $B(X^*)$ of the dual of a normed space X is compact in the weak-star topology.

Proof. For $x \in X$, set $D_x = \{\xi \in \mathbb{R} : |\xi| \leqslant \|x\|\}$ if X is a real space and $D_x = \{\xi \in \mathbb{C} : |\xi| \leqslant \|x\|\}$ if X is a complex space. Furthermore, let D be the product $\prod_{x \in X} D_x$ endowed with the product topology. Since D_x, a closed ball of the scalar field, is compact, Tychonov's theorem implies that D is a compact space.

Let us write B^* for $B(X^*)$ endowed with the weak-star topology, and define a map $\varphi : B^* \to D$ by setting, for $f \in B^*$,

$$\varphi(f) = (f(x))_{x \in X}.$$

The definitions of the product topology and the weak-star topology imply that φ is a continuous map of B^* into D, and it is clear that φ is one-to-one. Furthermore, again by the definitions of the product topology and the weak-star topology, φ^{-1} is a continuous map from $\varphi(B^*)$ onto B^*. Thus B^* is homeomorphic to the subset $\varphi(B^*)$ of D. Therefore, to complete the proof, it suffices to show that $\varphi(B^*)$ is a closed subset of the compact space D.

Let then $\xi = (\xi_x)_{x \in X} \in D$ be in the closure of $\varphi(B^*)$. Define a scalar function f on X by setting $f(x) = \xi_x$ for $x \in X$. We claim that f is a linear functional, i.e. $f \in X'$. To see this, let $x, y \in X$ and let α and β be scalars. Since $\xi \in \overline{\varphi(B^*)}$, for every $n \in \mathbb{N}$ there is a continuous linear functional $f_n \in B^*$ such that

$$|f(x) - f_n(x)| + |f(y) - f_n(y)| + |f(\alpha x + \beta y) - f_n(\alpha x + \beta y)| < \frac{1}{n}. \quad (4)$$

As $f_n(\alpha x + \beta y) = \alpha f_n(x) + \beta f_n(y)$, inequality (4) implies that

$$f(\alpha x + \beta y) = \alpha f(x) + \beta f(y),$$

i.e. $f \in X'$. Since $|f(x)| = |\xi_x| \leqslant \|x\|$ for every $x \in X$, f is not only a linear functional, but it is also continuous and $\|f\| \le 1$, i.e. $f \in B^*$. Finally, by the definition of f, we have $\xi = \varphi(f) \in \varphi(B^*)$, completing the proof. $\square$

As an immediate consequence of Alaoglu's theorem, we can see that a subspace of the Banach space of continuous functions on a compact Hausdorff space is the most general form of a normed space.

Theorem 4. Every normed space X is isometrically isomorphic to a subspace of $C(K)$, the Banach space of continuous functions on K, where K is $B(X^*)$ endowed with the weak-star topology. In particular, every normed space is isometrically isomorphic to a subspace of $C(K)$ for some compact Hausdorff space K.

Proof. Let X and K be as in the theorem; thus K is a compact Hausdorff space, namely $B(X^*)$ endowed with the weak-star topology. For $x \in X$ let f_x be the restriction of $\hat{x} \in X^{**}$ (defined by $\hat{x}(x^*) = x^*(x)$ for $x^* \in X^*$) to $K = B(X^*)$. By the definition of the weak-star topology, $f_x \in C(K)$; indeed, the weak-star topology on X^* is precisely the weakest topology in which every function $\hat{x}$ is continuous.

The map $x \mapsto f_x$ of X into $C(K)$ is clearly linear. Furthermore, by the Hahn–Banach theorem,

$$\|f_x\| = \sup\{|f_x(x^*)| : x^* \in K\}$$
$$= \sup\{|x^*(x)| : x^* \in B(X^*)\} = \|x\|,$$

and so the map $X \to C(K)$ is a linear isometry. $\square$

It is customary to write *w-topology* for the weak topology and *w*-topology* for the weak-star topology. Similarly, one may speak of *w*-open sets, *w**-closures, *w*-neighbourhoods, and so on, all taken in the appropriate spaces. Thus a *w*-open subset of X^* is a $\sigma(X^*, X^{**})$-open subset, and the *w**-closure of a subset of X^{**} is the closure of the set in the $\sigma(X^{**}, X^*)$-topology. As before, we consider X as a subspace of X^{**}, the isometric embedding being given by $x \mapsto \hat{x}$, where $\langle \hat{x}, x^* \rangle = \langle x^*, x \rangle$ for all $x^* \in X^*$.

In order to avoid some inessential complications, for the rest of the section we shall consider only real spaces.

The weak topology is weaker than the norm topology, so every *w*-closed set is also norm closed. The following result claims that for convex sets the converse is also true.

Theorem 5. A convex set C of a normed space X is closed iff it is w-closed.

Proof. Suppose that C is closed and $x_0 \notin C$. Then $d = d(x_0, C) > 0$. Set $D = \{x \in X : d(x_0, x) < d\}$. By the separation theorem (Theorem 3.13) there is a bounded linear functional $x^* \in X^*$ such that

$$s = \sup_{x \in C} \langle x^*, x \rangle \leq \inf_{x \in D} \langle x^*, x \rangle.$$

Then

$$U = \{x \in X : \langle x^*, x \rangle > s\}$$

is a w-neighbourhood of x_0 and $U \cap C = \varnothing$. This shows that C is also w-closed. As we remarked earlier, the converse is trivial. □

Theorem 6. $B(X^{**})$ is the w^*-closure of $B(X)$ in X^{**}.

Proof. It is easily checked that the closed unit ball $B(X^{**})$ is w^*-closed. Let B be the w^*-closure of $B(X)$. Since $B(X^{**})$ is w^*-closed and $B(X) \subset B(X^{**})$, we have $B \subset B(X^{**})$. Furthermore, as the w^*-closure of a convex set is convex, B is a convex set which is closed in the norm topology of X^{**}.

Suppose that, contrary to the assertion of the theorem, $B \neq B(X^{**})$. Then there is a point $x_0^{**} \in B(X^{**}) \backslash B$ so that, by the separation theorem (Theorem 3.13) there is a bounded linear functional x_0^{***} on X^{**} such that

$$\langle x_0^{***}, b \rangle \leq 1 < \langle x_0^{***}, x_0^{**} \rangle$$

for all $b \in B$. Let x_0^* be the restriction of x_0^{***} to the subspace X of X^{**}. Then

$$\langle x_0^*, x \rangle \leq 1$$

for all $x \in B(X)$ and so $\|x_0^*\| \leq 1$, contradicting

$$\langle x_0^*, x_0^{**} \rangle = \langle x_0^{***}, x_0^{**} \rangle > 1.$$ □

The results above have some beautiful consequences concerning reflexivity. Note that the w-topology on a normed space X is precisely the restriction to X of the w^*-topology on X^{**}.

Theorem 7. A normed space X is reflexive iff $B(X)$ is w-compact.

Proof. By Theorems 3 and 6, $B(X)$ is a w^*-dense subset of the w^*-compact set $B(X^{**})$. Since a subset of a compact Hausdorff space is compact iff it is closed, $B(X)$ is a w^*-compact subset of X^{**} iff $B(X) = B(X^{**})$. But this means precisely that $B(X)$ is w-compact iff $X = X^{**}$. $\qquad\square$

The next result can also be deduced from the Hahn–Banach theorem; now we are well equipped to give a very straightforward proof.

Corollary 8. A closed subspace of a reflexive space is reflexive.

Proof. Let Y be a closed subspace of a reflexive space X. Since $B(Y) = B(X) \cap Y$, the norm-closed convex set $B(Y) \subset X$ is $\sigma(X, X^*)$-compact. But, as we remarked in (2), the restriction of $\sigma(X, X^*)$ to Y is precisely $\sigma(Y, Y^*)$. Hence $B(Y)$ is w-compact and so, by Theorem 7, Y is reflexive. $\qquad\square$

Theorem 9. Let X be a Banach space. The following assertions are equivalent:
  (i) X is reflexive;
 (ii) $\sigma(X^*, X) = \sigma(X^*, X^{**})$, i.e. on X^* the w-topology and the w^*-topology coincide;
(iii) X^* is reflexive.

Proof. The implication (i) $\Rightarrow$ (ii) is trivial (and so is (i) $\Rightarrow$ (iii)). Suppose that (ii) holds. Since $B(X^*)$ is w^*-compact, it is also w-compact. Hence, by Theorem 7, X^* is reflexive. Thus (ii) $\Rightarrow$ (iii).

Finally, suppose that (iii) holds. As X is a Banach space, the ball $B(X)$ is norm closed in X^{**}. Therefore, by Theorem 5, $B(X)$ is w-closed, i.e. $\sigma(X^{**}, X^{***})$-closed. Since $X^{***} = X^*$, this means that $B(X)$ is w^*-closed in X^{**}. But, by Theorem 6, the w^*-closure of $B(X)$ is $B(X^{**})$, so $B(X) = B(X^{**})$. Hence (iii) $\Rightarrow$ (i). $\qquad\square$

If X if reflexive and $f \in X^*$ then the function $|f|$, being w^*-continuous on $B(X)$, attains its supremum there: for some $x_0 = B(X)$ we have

$$\|f\| = \sup\{|f(x)| : x \in B(X)\} = |f(x_0)|.$$

Equivalently, there is an $x_0 \in B(X)$ such that $\|f\| = f(x_0)$. (Of course, if $f \neq 0$ then x_0 is not only in the closed unit ball but also in the closed unit sphere $S(X)$.) It is a rather deep theorem of R. C. James that this property characterizes reflexive spaces: a Banach space X is reflexive iff every bounded linear functional attains its norm on $B(X)$.

Our next aim is to show that every Banach space comes rather close to having this property: the functionals which attain their suprema on the unit ball are dense in the dual space. (This property used to be called *subreflexivity*. As the result claims that every Banach space is subreflexive, the term has gone out of use.) In fact, we shall prove somewhat more.

Given a Banach space X, define

$$\Pi(X) = \{(x,f) : x \in S(X), f \in S(X^*), f(x) = 1\}.$$

Then $\Pi(X) \subset X \times X^*$ and, denoting the projection of $X \times X^*$ onto X by p_X and onto X^* by p_{X^*}, we have $p_X(\Pi(X)) = S(X)$. Furthermore, $p_{X^*}(\Pi(X))$ is precisely the set of functionals in $S(X^*)$ which attain their norms on $S(X)$.

For $\delta > 0$ let us write $\Pi_\delta(X)$ for the set of pairs $(x,f) \in S(X) \times S(X^*)$ such that f is *almost* 1 at x:

$$|f(x) - 1| < \delta.$$

We shall show that not only is $p_{X^*}(\Pi(X))$ dense in $S(X^*)$, which is precisely the subreflexivity of X, but also every point (x,f) of $\Pi_\delta(X)$ is close to some point (y,g) of $\Pi(X)$:

$$\|x - y\| + \|f - g\| < \varphi(\delta), \tag{5}$$

where $\varphi(\delta) \to 0$ as $\delta \to 0$.

The proof of this theorem is based on the following lemma, whose proof is left to the reader (see Exercise 13).

Lemma 10. Let X be a real normed space and let $f, g \in S(X^*)$ be such that

$$|g(x)| \leqslant \epsilon$$

whenever $x \in B(X)$ and $f(x) = 0$, where $0 < \epsilon < \frac{1}{2}$. Then either $\|f - g\| \leqslant 2\epsilon$ or $\|f + g\| \leqslant 2\epsilon$.   □

Theorem 11. (Bishop–Phelps–Bollobás theorem) Let X be a Banach space and let $x_0 \in S(X)$ and $f_0 \in S(X^*)$ be such that

$$|f_0(x_0) - 1| < \delta = \tfrac{1}{4}\epsilon^2,$$

where $0 < \epsilon < 1$. Then there exist $x_1 \in S(X)$ and $f_1 \in S(X^*)$ such that $f_1(x_1) = 1$, $\|f_0 - f_1\| \leqslant \epsilon$ and $\|x_0 - x_1\| \leqslant \epsilon$.

Proof. In the notation introduced before Lemma 10, the theorem claims that for every $(x_0, f_0) \in \Pi_\delta(X)$ there is an $(x_1, f_1) \in \Pi(X)$ such that $\|f_0 - f_1\| \leqslant \epsilon$ and $\|x_0 - x_1\| < \epsilon$; in particular, (5) holds with $\varphi(\delta) = 4\delta^{1/2}$, say.

As linear functionals are determined by their real parts and the map $f \mapsto \operatorname{Re} f$ is an isometry, we may assume that X is a real space. Let

$$ k = \left(1 + \frac{2}{\epsilon} \right) \Big/ f_0(x_0) $$

and define a partial order $\leqslant$ on $B = B(X)$ by setting $x \leqslant y$ if

$$ \|x - y\| \leqslant kf_0(y - x). $$

This is indeed a partial order, and a simple application of Zorn's lemma (see Exercise 16) shows that the set $B_0 = \{x \in B : x_0 \leqslant x\}$ has a maximal element, say x_1. Since $x_0 \leqslant x_1$,

$$ \|x_0 - x_1\| \leqslant kf_0(x_1 - x_0) \leqslant k\delta \leqslant \frac{1}{1-\delta}\left(1 + \frac{2}{\epsilon}\right)\delta = \frac{\frac{1}{2}\epsilon}{1 - \frac{1}{2}\epsilon} < \epsilon. $$

Set

$$ C = \frac{2}{\epsilon} B \cap \operatorname{Ker} f_0 $$

and let D be the convex hull of $B \cup C$:

$$ D = \operatorname{co}(B \cup D) = \{ ty + (1-t)z : y \in B \cup C, z \in B \cup C, 0 \leqslant t \leqslant 1 \} $$
$$ = \{ ty + (1-t)z : y \in B, z \in C, 0 \leqslant t \leqslant 1 \}. $$

Then D is a centrally symmetric convex set containing B and C.

We claim that $x_1 \notin \operatorname{Int} D$. Indeed, otherwise there are $0 < s < t < 1$, $y_1 \in B$ and $z_1 \in C$ such that

$$ x_1 = sy_1 + (t-s)z_1. $$

Since $f_0(x_0) > 0$ and $x \leqslant y$ implies that $f_0(x) \leqslant f_0(y)$, we see that

$$ f_0(x_0) \leqslant f_0(x_1) = sf_0(y_1) < f_0(y_1) $$

and so

$$ f_0(y_1 - x_1) = (1-s)f_0(y_1) > (1-s)f_0(x_0) > 0. \tag{6} $$

Furthermore,

$$y_1 - x_1 = (1-s) y_1 - (t-s) z_1,$$

and so

$$\|y_1 - x_1\| \leqslant (1-s) + (t-s)\frac{2}{\epsilon} < (1-s)\left(1 + \frac{2}{\epsilon}\right). \tag{7}$$

Inequalities (6) and (7) give

$$\|y_1 - x_1\| \leqslant (1-s)\left(1 + \frac{2}{\epsilon}\right) = \frac{1 + 2/\epsilon}{f_0(x_0)}(1-s)f_0(x_0) \leqslant kf_0(y_1 - x_1),$$

showing that $x_1 \leqslant y_1$. But, as x_1 is maximal, we have $x_1 = y_1$, contradicting (6), say. This proves that $x_1 \notin \operatorname{Int} D$, as claimed.

By the separation theorem (Theorem 3.13) there is a linear functional $f_1 \in X^*$ such that $f_1(x_1) = 1$ and $f_1(x) \leqslant 1$ for all $x \in D$. Then $\|f_1\| = 1$ and $f_1(x) \leqslant 1$ for all $x \in C$, and so $f_1(x) \leqslant \frac{1}{2}\epsilon$ for all $x \in B \cap \frac{1}{2}\epsilon \operatorname{Ker} f_0$. Hence, by Lemma 10,

$$\|f_0 - f_1\| \leqslant \epsilon \qquad \text{or} \qquad \|f_0 + f_1\| \leqslant \epsilon.$$

But

$$(f_0 + f_1)(x_1) = 1 + f_0(x_1) > 1 > \epsilon$$

and so we have $\|f_0 - f_1\| \leqslant \epsilon$. □

In Chapter 12 we shall give some applications of Theorem 11 to numerical ranges. Our last aim in this chapter is to prove another major result of functional analysis, the Krein–Milman theorem. Although in its proper generality this result has nothing to do with dual spaces and weak topologies, we present it here since it has some natural applications concerning dual spaces.

Our proof of the Krein–Milman theorem is based on another equivalent form of the axiom of choice, namely on the well-ordering principle, enabling us to apply transfinite induction.

An *ordered set* is a set with a linear (total) order. An ordered set $(S, \leqslant)$ is said to be *well ordered* if every (non-empty) subset T of S has a smallest element, i.e. an element $t_0 \in T$ such that $t_0 \leqslant t$ for all $t \in T$.

Lemma 12. (Well-ordering principle) Every set can be well ordered.

Proof. A subset B of an ordered set $(A, \leqslant)$ is said to be an *initial segment* if $a \in A$, $b \in B$ and $a \leqslant b$ imply that $a \in B$.

Let S be a set and let $P = \{(S_\gamma, \leqslant_\gamma) : \gamma \in \Gamma\}$ be the set of all well ordered sets such that S_γ is a subset of S. For $(S_\gamma, \leqslant_\gamma)$ and $(S_\delta, \leqslant_\delta)$ in

P set

$$(S_\gamma, \leqslant_\gamma) \leqslant (S_\delta, \leqslant_\delta)$$

if S_γ is an initial segment of S_δ and on S_γ the two orderings coincide. Clearly $(P, \leqslant)$ is a partially ordered set.

It is easily seen that in $(P, \leqslant)$ every chain has an upper bound. Indeed, let $\{(S_\gamma, \leqslant_\gamma) : \gamma \in \Gamma_0\}$ be a chain. Set $S' = \bigcup_{\gamma \in \Gamma_0} S_\gamma$ and for $a, b \in S'$ set $a \leqslant' b$ if $a, b \in S_\gamma$ and $a \leqslant_\gamma b$ for some $\gamma \in \Gamma_0$. Then $\leqslant'$ is an order on S'. Furthermore, if $T' \subset S'$ and $t' \in T'$ then $t' \in S_\gamma$ for some $\gamma \in \Gamma$, and the smallest element of $T' \cap S_\gamma$ in $(S_\gamma, \leqslant_\gamma)$ is also a smallest element of T' in $(S, \leqslant')$.

Since every chain has an upper bound, by Zorn's lemma from Chapter 3, $(P, \leqslant)$ has a maximal element $(S_0, \leqslant_0)$. To complete the proof, all we have to show is that $S_0 = S$. Suppose that this is not so, say $s_1 \in S \backslash S_0$. Extend the order $\leqslant_0$ to an order $\leqslant_1$ on $S_1 = S_0 \cup \{s_1\}$ by setting $s \leqslant_1 s_1$ for all $s \in S_0$. Then $(S_1, \leqslant_1)$ is a well ordered set and $(S_0, \leqslant_0) < (S_1, \leqslant_1)$, contradicting the maximality of $(S_0, \leqslant_0)$. ☐

Let K be a convex subset of a vector space. A point in K is said to be an *extreme point* of K if it is not in the interior of any line segment entirely in K. Thus $a \in K$ is an extreme point of K if whenever $b, c \in K$, $0 < t < 1$ and $a = tb + (1-t)c$, we have $a = b = c$. The set of extreme points of K is denoted by $\text{Ext}\, K$.

Recall from chapter 3 that the *convex hull* of a subset S of a vector space V is

$$\text{co}\, S = \left\{ \sum_{i=1}^{n} t_i x_i : x_i \in S, t_i \geqslant 0 \ (i = 1, \dots, n), \sum_{i=1}^{n} t_i = 1 \ (n = 1, 2, \dots) \right\}.$$

If S is in a *normed* space then $\overline{\text{co}}\, S$, the *closed convex hull* of S, is the closure of $\text{co}\, S$; by Theorem 3.14, $\overline{\text{co}}\, S$ is the intersection of all closed half-spaces containing S.

In its full generality, the Krein–Milman theorem concerns locally convex topological vector spaces. As we do not study these spaces, we shall be satisfied with a somewhat artificial form of this result. Given a vector space V with dual V', for $F \subset V'$ and $S \subset V$ the *F-closed convex hull* of S is

$$\overline{\text{co}}_F S = \bigcap_{f \in F} \{x \in V : f(x) \leqslant \sup_{y \in S} f(y)\}.$$

For the rest of the chapter, we say that S is *F-convex* if $\overline{\text{co}}_F S = S$. By

Theorem 3.14, if X is a normed space then $\overline{co}\,S = \overline{co}_{X^*}S$ for all $S \subset X$; a set X is convex and closed (in the norm topology) iff it is X^*-convex.

Theorem 13. (Krein–Milman theorem) Let τ be a Hausdorff topology on a vector space V and let $F \subset V'$. Suppose that F separates the points of V (i.e. if $f(v) = 0$ for all $f \in F$ then $v = 0$) and each $f \in F$ is τ-continuous. Let K be a non-empty τ-compact F-convex subset of V. Then $\mathrm{Ext}\,K \neq \varnothing$ and $K = \overline{co}_F\mathrm{Ext}\,K$.

Proof. Let $\{f_\gamma : \gamma \in \Gamma\}$ be a well ordering of F. Then there is a function $s : \Gamma \to \mathbb{R}$, $\gamma \mapsto s_\gamma$, such that for all $\gamma \in \Gamma$ we have

$$s_\gamma = \max\{f_\gamma(x) : x \in K,\ f_\delta(x) = s_\delta \text{ for } \delta < \gamma\}. \tag{8}$$

This is immediate from the fact that F (i.e. Γ) is well ordered and if s_δ is determined for all $\delta \in \Gamma$ preceding $\gamma \in \Gamma$ then

$$K \cap \bigcap_{\delta < \gamma} f_\delta^{-1}(s_\delta)$$

is a non-empty compact set on which the continuous function f_γ attains its supremum, s_γ.

The existence of the function s implies that there is a point $a \in K$ such that $f_\gamma(a) = s_\gamma$ for all $\gamma \in \Gamma$. This point a is an extreme point of K. Indeed, if $a = tb + (1-t)c$ for some $0 < t < 1$ and $b, c \in K$, then (8) implies that $f_\gamma(a) = f_\gamma(b) = f_\gamma(c)$ for all $\gamma \in \Gamma$. As F separates the points of V, we have $a = b = c$. Thus $\mathrm{Ext}\,K \neq \varnothing$.

Set $K' = \overline{co}_F\mathrm{Ext}\,K$. Then $K' \subset K$ since K is F-convex. To complete the proof, we have to show that, in fact, $K' = K$. Suppose that this is not so. Then there is a functional $f_0 \in F$ such that

$$\max_{x \in K'} f_0(x) < \max_{x \in K} f_0(x). \tag{9}$$

Choosing a well ordering of F in which f_0 is the first element, the point a produced by the procedure above is such that $f_0(a) = \max_{x \in K}f_0(x)$. But $a \in K'$, contradicting (9). $\qquad\square$

Corollary 14. Let K be a compact convex subset of a Banach space. Then $K = \overline{co}\,\mathrm{Ext}\,K$. $\qquad\square$

Corollary 15. Let X be a normed space. Then $B(X^*) = \overline{co}\,\mathrm{Ext}\,B(X^*)$.

Proof. By Alaoglu's theorem, $B(X^*)$ is w^*-compact. Furthermore, it is X-convex since if $f_0 \in X^*$ and $f_0 \notin B(X^*)$ then $f_0(x_0) > 1$ for some $x_0 \in B(X)$ and $f(x_0) \leqslant 1$ for all $x \in B(X)$. $\qquad\square$

The last result implies that $B(X^*)$ has 'many" extreme points: sufficiently many to guarantee a 'large" closed convex hull, namely $B(X^*)$. Occasionally this fact is sufficient to ensure that a given space is not a dual space. For example, it is easily seen that $\text{Ext} B(c_0) = \varnothing$; so c_0 is not a dual space. Indeed, let $x = (x_n)_1^\infty \in B(c_0)$. Then $|x_k| < \frac{1}{2}$ for some k and so, with $y_n = z_n = x_n$ for $n \neq k$, $y_k = x_k + \frac{1}{2}$ and $z_k = x_k - \frac{1}{2}$, we have $y = (y_n)_1^\infty$, $z = (z_n)_1^\infty \in B(c_0)$ and $x = \frac{1}{2}(y+z)$, so that $x \notin \text{Ext} B(c_0)$. Thus $\text{Ext} B(c_0)$ is indeed empty.

The Krein–Milman theorem is one of the fundamental results of functional analysis, leading to deep theorems in operator theory and abstract harmonic analysis, like the Gelfand–Naimark theorem concerning irreducible *-representations of C^*-algebras, and the Gelfand–Raĭkov theorem about unitary representations of locally compact groups. But these results will not be presented in this book. We shall return to compact convex sets in Chapters 13–16, when we study compact operators and fixed-point theorems. However, our next aim is to study the 'nicest' Banach spaces, namely Hilbert spaces.

Exercises

1. Prove identity (2), i.e. show that if Y is a subspace of a normed space X then the weak topology on Y is precisely the topology induced on Y by the weak topology on X.

2. Let Y be a closed subspace of a reflexive space X, and let $x_0 \in X$. Show that the function $x \mapsto \|x - x_0\|$ is weakly lower semicontinuous, i.e. it is lower semicontinuous in the w-topology on X. Deduce that the distance of x_0 from Y is attained, i.e. there is a point $y_0 \in Y$ such that

$$\|x_0 - y_0\| = d(x_0, Y) = \inf\{\|x_0 - y\| : y \in Y\}.$$

3. Show that if Y is a closed subspace of a reflexive space X then X/Y is reflexive.

4. Show that

$$d(x, y) = \sum_{n=1}^\infty 2^{-n} |x_n - y_n|$$

defines a metric on l_∞, and the restriction of this metric to $B(l_\infty)$ induces the weak-star topology on $B(l_\infty)$.

5. Let X be an infinite-dimensional normed space. Show that the w-closure of $S(X) = \{x \in X : \|x\| = 1\}$ is $B(X) = \{x \in X : \|x\| \leq 1\}$.

6. Let (x_n) be a sequence in l_p $(1 < p < \infty)$ weakly converging to $x_0 \in l_p$ (thus $x_n \to x_0$ in the w-topology, i.e. $\langle x_n, x^* \rangle \to \langle x_0, x^* \rangle$ for every $x^* \in l_p^* = l_q$). Prove that if $\|x_n\| \to \|x_0\|$ then $\|x_n - x_0\| \to 0$.
7. Show that a sequence $(x_n) \subset l_p$ $(1 < p < \infty)$ converges weakly to 0 iff (x_n) is bounded (i.e. $\|x_n\| \leqslant K$ for some K and all n) and $\lim_{n\to\infty} x_i^{(n)} = 0$ for every i, where $x_n = (x_1^{(n)}, x_2^{(n)}, \ldots)$.
8. Let $x_n = (x_1^{(n)}, x_2^{(n)}, \ldots) \in l_1$ $(n = 1, 2, \ldots)$. Show that $\langle x_n, y \rangle \to 0$ for every $y \in c_0$ iff $(x_n)_1^\infty$ is bounded and $\lim_{n\to\infty} x_i^{(n)} = 0$ for every i.
9. Formulate and prove assertions analogous to the previous two exercises for the spaces $L_p(0,1)$ and $L_q(0,1)$.
10. Let X be an infinite-dimensional normed space. Show that the norm function $\|x\|$ $(x \in X)$ is not weakly continuous at any point of X.
11. Let $(e_n)_1^\infty$ be the standard basis of l_2 and let

$$A = \{e_m + m e_n : 1 \leqslant m < n\}.$$

Prove that 0 is in the w-closure of A but no sequence in A converges weakly to 0.
12. Let X be a Banach space, $f \in X^*$ and $M = \operatorname{Ker} f$. Prove that for every $x \in X$ there is a point of M nearest to x iff f attains its norm on $B(X)$.
13. Prove Lemma 10: if X is a real normed space and $f, g \in S(X^*)$ are such that $(\operatorname{Ker} f) \cap B(X) \subset g^{-1}[-\epsilon, \epsilon]$ for some $0 < \epsilon < \frac{1}{2}$, then $\|f - g\| \leqslant 2\epsilon$ or $\|f + g\| \leqslant 2\epsilon$. [HINT: If $f \neq \pm g$ then there are $a, b \in X$ such that $f(a) = g(b) = 1$ and $f(b) = g(a) = 0$. Set $M = \operatorname{lin}\{a, b\}$ and $N = \operatorname{Ker} f \cap \operatorname{Ker} g$, and denote by D the projection of $B(X)$ into M parallel to N. Let $\|\cdot\|'$ be the norm on M with unit ball D, and note that with $f_0 = f|M$ and $g_0 = g|M$ we have $\|\alpha f_0 + \beta g_0\|' = \|\alpha f + \beta g\|$ for all $\alpha, \beta \in \mathbb{R}$. This shows that it suffices to prove the lemma in the case when $X = M$, i.e. when X is 2-dimensional. Use a simple geometric argument to complete the proof.]
14. Show that the assertion of Lemma 10 is best possible: for $0 < \epsilon < \frac{1}{2}$, there are functionals $f, g \in S(X^*)$ such that $(\operatorname{Ker} f) \cap B(X) \subset g^{-1}[-\epsilon, \epsilon]$, $\|f + g\| \geqslant 1$ and $\|f - g\| = 2\epsilon$.
15. Show that Zorn's lemma indeed implies the existence of a maximal element $x_1 \geqslant x_0$ in the proof of Theorem 11.
16. Check that Theorem 11 can be strengthened a little: if $|f_0(x_0) - 1| < \frac{1}{2}\epsilon^2$, where $0 < \epsilon < \frac{1}{2}$ then we may demand

$\|f_0 - f_1\| \leq \epsilon$ and $\|x_0 - x_1\| < \epsilon + \epsilon^2$. (This variant of the result is essentially best possible.)

17. Let $K \subset \mathbb{R}^n$ be a compact convex set. Show that if $n = 2$ then $\text{Ext}\, K$ is compact, but this need not be the case if $n \geq 3$.

18. Show that $B(l_1) = \overline{\text{co}}\, \text{Ext}\, B(l_1)$.

19. What are the extreme points of $C^{\mathbb{R}}[0, 1]$?

20. Let L be a locally compact Hausdorff space, which is not compact. Show that the closed convex set $B(C_0^{\mathbb{R}}(L))$ has no extreme points.

21. Let X be an infinite-dimensional Banach space whose unit ball has only finitely many extreme points. Show that X is not a dual space, i.e. there is no normed space Y such that $X = Y^*$.

22. Show that $C^{\mathbb{R}}[0, 1]$ is not a dual space.

Notes

Tychonov's theorem is from A. Tychonoff, *Über ein Funktionenraum*, Mathematische Annalen, **111** (1935), 762–66; somewhat earlier, an incomplete version of the result appeared in A. Tychonoff, *Über die topologische Erweiterung von Räumen*, Mathematische Annalen, **102** (1929), 544–61. (The modern transliteration of the name 'Tychonov' is closer to the Russian original than the one used in German before the second world war. Similarly, the old-fashioned usage of Markoff, Tschebischeff, Egoroff, etc., is due to German influence.) A proof of the thoerem can be found in almost any good book on general topology, for example, in J. L. Kelley, *General Topology*, Van Nostrand, Princeton, N.J., 1955, xiv + 298 pp.

Theorem 3 is from L. Alaoglu, *Weak topologies of normed linear spaces*, Annals of Math., **41** (1940), 252–67; in fact, the result was announced two years earlier in Leonidas Alaoglu, *Weak convergence of linear functionals*, Bull. American Mathematical Society, **44** (1938), p. 196.

The theorem of James mentioned after Theorem 9 is from R. C. James, *Characterizations of reflexivity*, Studia Math., **23** (1964), 205–16. The original form of Theorem 11, stating that for a Banach space X the set of functionals attaining their norms on $S(X)$ is dense in X^*, is from E. Bishop and R. R. Phelps, *A proof that every Banach space is subreflexive*, Bull. Amer. Math. Soc., **67** (1961), 97–8; the form we gave it in is from B. Bollobás, *An extension to the theorem of Bishop and Phelps*, Bull. London Math. Soc., **2** (1970), 181–2. Our presentation followed that given in F. F. Bonsall and J. Duncan, *Numerical Ranges II*, London Math. Soc. Lecture Note Series, vol. 10, Cambridge University Press, 1973, vii + 179 pp.

9. EUCLIDEAN SPACES AND HILBERT SPACES

Having studied general normed spaces and Banach spaces, in the next two chapters we shall look at the 'nicest' examples of these spaces, namely Euclidean spaces and Hilbert spaces. These spaces are the natural generalizations of the n-dimensional Euclidean space l_2^n.

A *hermitian form* on a complex vector space V is a map $f : V \times V \to \mathbb{C}$ such that for all $x, y, z \in V$ and $\lambda, \mu \in \mathbb{C}$ we have

$$\text{(i)} \quad f(\lambda x + \mu y, z) = \lambda f(x, z) + \mu f(y, z)$$

$$\text{(ii)} \quad f(y, x) = \overline{f(x, y)} \qquad \text{for all } x, y \in V.$$

Similarly, a *hermitian form* on a real vector space V is a map $f : V \times V \to \mathbb{R}$ satisfying (i) and (ii). In that case (ii) says simply that f is *symmetric*, so a real hermitian form is just a symmetric bilinear form. If f is a hermitian form on a complex vector space V then $g = \text{Re} f$ is a real hermitian form on $V_{\mathbb{R}}$, the space V considered as a real vector space. Conversely, for every real hermitian form g on $V_{\mathbb{R}}$ there is precisely one complex hermitian form f on V with $g = \text{Re} f$ (see Exercise 1).

Let f be a hermitian form on a real or complex vector space V. Note that $f(x, x)$ is real for all $x \in V$ and $f(\lambda x, \lambda x) = |\lambda|^2 f(x, x)$ for all $x \in V$ and scalar λ. We call f *positive* if, in addition to (i) and (ii), it satisfies

$$\text{(iii)} \quad f(x, x) \geq 0 \qquad \text{for all } x \in V.$$

Two vectors x and y are said to be *orthogonal with respect to f* if $f(x, y) = 0$; orthogonality is sometimes expressed by writing $x \perp y$. A vector orthogonal to itself is said to be *isotropic*. We call f *degenerate* if some non-zero vector is orthogonal to the entire space; otherwise the form is *non-degenerate*.

Theorem 1. Let f be a positive hermitian form on V. Then for all $x, y \in V$ we have

$$|f(x, y)|^2 \leqslant f(x, x) f(y, y) \tag{1}$$

and

$$f(x + y, x + y)^{1/2} \leqslant f(x, x)^{1/2} + f(y, y)^{1/2}. \tag{2}$$

Furthermore, f is non-degenerate iff 0 is the only isotropic vector with respect to f.

Proof. In proving (1), we may replace x by λx if $|\lambda| = 1$ so we may assume that $f(x, y)$ is real. Then for all $t \in \mathbb{R}$ we have

$$0 \leqslant f(tx + y, tx + y) = t^2 f(x, x) + 2tf(x, y) + f(y, y)$$

so (1) follows.

To see (2), note that, by (1),

$$f(x + y, x + y) \leqslant f(x, x) + 2|f(x, y)| + f(y, y)$$

$$\leqslant f(x, x) + 2[f(x, x) f(y, y)]^{1/2} + f(y, y)$$

$$= [f(x, x)^{1/2} + f(y, y)^{1/2}]^2.$$

Finally, if x is isotropic then (1) implies that x is orthogonal to every vector in V. $\qquad\square$

Inequality (1) is, of course, just the *Cauchy–Schwarz inequality*; inequality (2) is a variant of *Minkowski's inequality* for $p = 2$.

In view of Theorem 1, a positive hermitian form is non-degenerate iff it is a *positive-definite hermitian form*, i.e. a hermitian form f such that if $x \neq 0$ then $f(x, x) > 0$.

An *inner-product space* is a vector space V together with a positive definite hermitian form on V. This positive-definite hermitian form is said to be the *inner product* or *scalar product* on V and we shall write it as $(\cdot, \cdot)$ (and, occasionally, as $\langle \cdot, \cdot \rangle$). Thus $(\cdot, \cdot)$ is such that for all $x, y, z \in V$ and for all scalars λ and μ we have

(i) $(\lambda x + \mu y, z) = \lambda(x, z) + \mu(y, z)$;

(ii) $(y, x) = \overline{(x, y)}$;

(iii) $(x, x) \geqslant 0$, with equality iff $x = 0$.

More often than not, it will not matter whether our inner-product space is a *complex* or a *real* inner-product space. As the complex case tends to look a little more complicated, usually we shall work with complex spaces. By Theorem 1, if $(\cdot, \cdot)$ is an inner product on a vector space V then $\|x\| = (x,x)^{1/2}$ is a norm on V. A normed space is said to be a *Euclidean space* or a *pre-Hilbert space* if its norm can be derived from an inner product. A complete Euclidean space is called a *Hilbert space*. Clearly every subspace of a Euclidean space is a Euclidean space and every closed subspace of a Hilbert space is a Hilbert space.

The Cauchy–Schwarz inequality states that $|(x,y)| \leqslant \|x\|\|y\|$; in particular, the inner product is jointly continuous in the induced norm. The inner product defining a norm can easily be recovered from the norm. Indeed, we have the following *polarization identities*:

$$4(x,y) = \|x+y\|^2 - \|x-y\|^2 + i\|x+iy\|^2 - i\|x-iy\|^2 \tag{3}$$

if the space is complex, and

$$2(x,y) = \|x+y\|^2 - \|x\|^2 - \|y\|^2 = \tfrac{1}{2}\{\|x+y\|^2 - \|x-y\|^2\} \tag{4}$$

in the real case. Therefore in a Euclidean space we may, and often shall, use the inner product defining the norm. For this reason, we use the terms 'Euclidean space' and 'inner-product space' interchangeably, and we may talk of *orthogonal* vectors in a Euclidean space.

The complex polarization identity (3) has the following simple extension. If T is a linear operator (not necessarily bounded) on a complex Euclidean space then

$$4(Tx,y) = (T(x+y), x+y) - (T(x-y), x-y)$$
$$+ i(T(x+iy), x+iy) - i(T(x-iy), x-iy). \tag{3'}$$

This implies the following result.

Theorem 2. Let E be a complex Euclidean space and let $T \in \mathcal{L}(E)$ be such that $(Tx,x) = 0$ for all $x \in E$. Then $T = 0$.

Proof. By (3') we have $(Tx,y) = 0$ for all $x, y \in E$. In particular, $\|Tx\|^2 = (Tx, Tx) = 0$ for all $x \in E$, so $T = 0$.   □

It is worth pointing out that Theorem 2 cannot be extended to real Euclidean spaces (see Exercise 2).

In a Euclidean space, the theorem of Pythagoras holds and so does the parallelogram law.

Theorem 3. Let E be a Euclidean space. If $x_1, \ldots, x_n$ are pairwise orthogonal vectors then

$$\left\|\sum_{i=1}^{n} x_i\right\|^2 = \sum_{i=1}^{n} \|x_i\|^2. \tag{5}$$

Furthermore, if $x, y \in E$ then

$$\|x+y\|^2 + \|x-y\|^2 = 2\|x\|^2 + 2\|y\|^2. \tag{6}$$

Proof. Both (5), the *Pythagorean theorem*, and (6), the *parallelogram law*, are immediate upon expanding the sides as sums of products. To spell it out,

$$\left\|\sum_{i=1}^{n} x_i\right\|^2 = \left(\sum_{i=1}^{n} x_i, \sum_{i=1}^{n} x_i\right) = \sum_{i=1}^{n} (x_i, x_i) + \sum_{i \ne j} (x_i, x_j)$$

$$= \sum_{i=1}^{n} (x_i, x_i) = \sum_{i=1}^{n} \|x_i\|^2$$

and

$$\|x+y\|^2 + \|x-y\|^2 = (x+y, x+y) - (x-y, x-y) = 2\|x\|^2 + 2\|y\|^2. \quad \square$$

In fact, the parallelogram law characterizes Euclidean spaces (see Exercise 3). This shows, in particular, that a normed space is Euclidean iff all its two-dimensional subspaces are Euclidean. Furthermore, a complex normed space is Euclidean iff it is Euclidean when considered as a real normed space.

The last assertion is easily justified without the above characterization of Euclidean spaces. Indeed, if E is an Euclidean space then $\text{Re}(x, y)$ is a real inner product on the underlying real vector space of E and it defines the same norm on E. Conversely, if V is a complex vector space and $\langle \cdot, \cdot \rangle$ is a *real* inner product on V (i.e. $\langle \cdot, \cdot \rangle$ is an inner product on $V_{\mathbb{R}}$) such that the induced norm $\|\cdot\|$ satisfies $\|\lambda x\| = |\lambda| \|x\|$ for all $x \in V$ and $\lambda \in \mathbb{C}$ then

$$(x, y) = \langle x, y \rangle - i\langle ix, y \rangle = \langle x, y \rangle + i\langle x, iy \rangle$$

is a complex inner product on V defining the original norm $\|\cdot\|$ and satisfying $\text{Re}(x, y) = \langle x, y \rangle$.

Examples 4. (i) Clearly, $(x, y) = \sum_{i=1}^{n} x_i \bar{y}_i$ is an inner product on l_2^n and so l_2^n is an n-dimensional Hilbert space.

(ii) Also, $(x, y) = \sum_{i=1}^{\infty} x_i \bar{y}_i$ is an inner product on l_2 and so l_2 is a separable infinite-dimensional Hilbert space.

(iii) Let E be the vector space of all eventually zero sequences of complex numbers (i.e. $x = (x_i)_1^{\infty}$ belongs to E if $x_i = 0$ whenever i is sufficiently large), with inner product $(x, y) = \sum_{i=1}^{\infty} x_i \bar{y}_i$. Then E is a dense subspace of l_2; it is an incomplete Euclidean space.

(iv) It is immediate that

$$(f, g) = \int_a^b f(t) \overline{g(t)} \, dt$$

is an inner product on the vector space $C([a, b])$; the norm defined is the l_2-norm

$$\|f\|_2 = \left(\int_a^b |f(t)|^2 \, dt \right)^{1/2}$$

and not the uniform norm. This is an incomplete Euclidean space (see Exercise 14).

(v) Also,

$$(f, g) = \int_a^b f(t) \overline{g(t)} \, dt$$

is the inner product on

$$L_2(0, 1) = \left\{ f \in \mathbb{C}^{[0, 1]} : f \text{ is measurable and } \int_0^1 |f(t)|^2 \, dt < \infty \right\},$$

defining the L_2-norm

$$\|f\|_2 = \left(\int_0^1 |f(t)|^2 \, dt \right)^{1/2}.$$

With this norm $L_2(0, 1)$ is a Hilbert space.

(vi) Let $V = \mathbb{C}^n$, let $A = (a_{ij})$ be an $n \times n$ complex matrix and, for $x = (x_i)_1^n$ and $y = (y_i)_1^n$ in V define

$$f(x, y) = \sum_{i, j = 1}^{n} a_{ij} x_i \bar{y}_j.$$

Then f is a hermitian form if $a_{ji} = \bar{a}_{ij}$ for all i, j. Also, every hermitian form on V can be obtained in this way. □

Theorem 5. The completion of a Euclidean space is a Hilbert space.

Proof. The assertion is that the inner product $(\cdot, \cdot)$ on a Euclidean space E can be extended to the completion $\tilde{E}$ of E. Let (x_n) and (y_n) be Cauchy sequences in E and let $\tilde{x}$ and $\tilde{y}$ be their equivalence classes in $\tilde{E}$. Define

$$\langle \bar{x}, \bar{y} \rangle = \lim_{n \to \infty} (x_n, y_n).$$

Since

$$|(x_n, y_n) - (x_m, y_m)| \leq |(x_n - x_m, y_n)| + |(x_m, y_n - y_m)|$$

$$\leq \|x_n - x_m\| \|y_n\| + \|x_m\| \|y_n - y_m\|,$$

the limit exists. It is easily seen to depend only on $\bar{x}$ and $\bar{y}$ and not on the particular representatives (x_n), (y_n). Furthermore, $\langle \cdot, \cdot \rangle$ is an inner product on $\bar{E}$, extending E and defining precisely the norm on the completion $\bar{E}$. $\qquad\square$

Let E be a Euclidean space. For $x \in E$, write $x^\perp$ for the set of vectors orthogonal to x:

$$x^\perp = \{y \in E : x \perp y\} = \{y \in E : (x, y) = 0\}.$$

Clearly $x^\perp$ is a closed subspace of E and therefore so is

$$S^\perp = \{y \in E : x \perp y \text{ for all } x \in S\} = \bigcap_{x \in S} x^\perp$$

for every subset S of E. If F is the closed linear span of S, i.e. $F = \overline{\text{lin}}\, S$, then $S^\perp = F^\perp$ and $(S^\perp)^\perp \supset F$.

Call two subspaces F_1 and F_2 *orthogonal* if $x_1 \perp x_2$ for all $x_1 \in F_1$ and $x_2 \in F_2$. If F is a subspace of E then F and $F^\perp$ are orthogonal subspaces and $F \cap F^\perp = \{0\}$ since if $x \in F \cap F^\perp$ then $x \perp x$ and so $x = 0$. If $x \in F$ and $y \in F^\perp$ then $\|x + y\|^2 = \|x\|^2 + \|y\|^2$, so the projections $x + y \mapsto x$ and $x + y \mapsto y$ are both bounded. Thus $F + F^\perp$ is the *orthogonal direct sum* of F and $F^\perp$. In general, $F + F^\perp$ need not be the entire space, not even when F is closed (see Exercise 6). However, when F is complete, this is the case.

Theorem 6. Let F be a complete subspace of a Euclidean space E. Then E is the orthogonal direct sum of F and $F^\perp$: every $x \in E$ has a unique representation in the form

$$x = x_1 + x_2 \qquad (x_1 \in F, \, x_2 \in F^\perp).$$

Furthermore, if $y \in F$ and $y \neq x_1$ then

$$\|x - y\| > \|x - x_1\| = \|x_2\|. \tag{7}$$

Proof. Let $x \in E$ and set $d = d(x, F)$. What characterizes x_1? By (7), it has to be the unique vector in F nearest to x, precisely at distance d.

Let then $y_n \in F$ be such that

$$\|x-y_n\|^2 < d^2 + \frac{1}{n}.$$

By identity (6), the parallelogram law,

$$\|y_n - y_m\|^2 = 2\|x-y_n\|^2 + 2\|x-y_m\|^2 - \|2x-y_n-y_m\|^2$$
$$\leq \frac{2}{n} + \frac{2}{m},$$

and so $(y_n)_1^\infty$ is a Cauchy sequence. Since F is complete, $y_n \to x_1$ for some $x_1 \in F$. Then $\|x-x_1\| = d$.

Let us show that $x_2 = x - x_1$ is orthogonal to F. Suppose this is not so. Then F has an element, say y, such that $(x_2, y) \neq 0$. Replacing y by $z = (x_2, y)y$ we have $(x_2, z) \in \mathbb{R}$ and $(x_2, z) > 0$. But then for sufficiently small $\epsilon > 0$ we have

$$\|x-(x_1+\epsilon z)\|^2 = \|x_2 - \epsilon z\|^2$$
$$= d^2 - 2(x_2, z)\epsilon + \|z\|^2\epsilon^2 < d^2,$$

contradicting $d(x, F) = d$.

Inequality (7) is immediate from identity (5), the Pythagorean theorem. If $y \in F$ and $x_1 - y \neq 0$ then, as $(x_1 - y) \perp x_2$, we have

$$\|x-y\|^2 = \|x_1 + x_2 - y\|^2 = \|x_1 - y\|^2 + \|x_2\|^2 > \|x_2\|^2 = d^2. \quad \square$$

In view of Theorem 6, if F is a complete subspace of a Euclidean space E then we call $F^\perp$ the *orthogonal complement* of F in E. The map $P_F: E = F + F^\perp \to E$ defined by $x = x_1 + x_2 \mapsto x_1$ is called the *orthogonal projection* onto F; we have just shown that there is such a map.

Corollary 7. Let F be a complete subspace of a Euclidean space E. Then there is a unique operator $P_F \in \mathscr{B}(E)$ such that

$$P_F(x) = \begin{cases} x & \text{for } x \in F, \\ 0 & \text{for } x \in F^\perp. \end{cases}$$

Furthermore,

$$\operatorname{Im} P_F = F, \quad \operatorname{Ker} P_F = F^\perp, \quad P_F^2 = P_F, \quad (I-P_F)^2 = I-P_F$$

and if $x, y \in E$ then

$$(P_F x, y) = (P_F x, P_F y) = (x, P_F y).$$

If $F \neq (0)$ then $\|P_F\| = 1$ and if $F \neq E$ then $\|I - P_F\| = 1$. $\quad \square$

Corollary 8. Let H be a Hilbert space, let $S \subset H$ and let M be the closed linear span of S. Then $(S^\perp)^\perp = (M^\perp)^\perp = M$.

Proof. As $H = M + M^\perp = (M^\perp)^\perp + M^\perp$ and $M \subset (M^\perp)^\perp$, we have

$$M = (M^\perp)^\perp = (S^\perp)^\perp. \qquad \square$$

Given a Hilbert space H and a vector $x_0 \in H$, the function $f: H \to \mathbb{C}$ defined by $f(x) = (x, x_0)$ is a bounded linear functional of norm $\|x_0\|$. Indeed, f is clearly linear, and $|f(x)| = |(x, x_0)| \leqslant \|x_0\| \|x\|$ and so $\|f\| \leqslant \|x_0\|$. Furthermore, $|f(x_0)| = \|x_0\|^2$ and so $\|f\| \geqslant \|x_0\|$. The existence of an orthogonal decomposition implies that *all* bounded linear functionals on H can be obtained in this way.

Theorem 9. (Riesz representation theorem) Let f be a bounded linear functional on a Hilbert space H. Then there is a unique vector $x_0 \in H$ such that $f(x) = (x, x_0)$ for all $x \in H$. Furthermore, $\|f\| = \|x_0\|$.

Proof. We may assume that $f \neq 0$. Then $M = \mathrm{Ker}\, f$ is a closed one-codimensional subspace of H. Consequently $M^\perp$ is one-dimensional, say $M^\perp = \{\lambda x_1 : \lambda \in \mathbb{C}\}$, where $x_1 \in H$ and $\|x_1\| = 1$.

Put $x_0 = \overline{f(x_1)} x_1$. Every vector $x \in H$ has a unique representation in the form

$$x = y + \lambda x_1, \qquad (y \in M; \lambda \in \mathbb{C}).$$

Then

$$(x, x_0) = (y + \lambda x_1, \overline{f(x_1)} x_1) = (\lambda x_1, \overline{f(x_1)} x_1)$$

$$= \lambda f(x_1) = f(\lambda x_1) = f(y + \lambda x_1) = f(x).$$

To see the uniqueness of x_0, note that if $(x, x_0) = (x, x_0') = 0$ for all $x \in H$, then

$$\|x_0 - x_0'\|^2 = (x_0 - x_0', x_0 - x_0') = 0.$$

The assertion $\|f\| = \|x_0\|$ was shown immediately before the theorem. $\qquad \square$

Corollary 10. Let H be a Hilbert space. For $y \in H$ let $f_y \in H^*$ be defined by $f_y(x) = (x, y)(x \in H)$. Then the map $H \to H^*$ defined by $y \mapsto f_y$ is an isometric anti-isomorphism between H and H^*, i.e. if $y \mapsto f$ and $z \mapsto g$ then $\lambda y + \mu g \mapsto \overline{\lambda} f + \overline{\mu} g$. If H is a real Hilbert space then the map is an isometric isomorphism between H and H^*. $\qquad \square$

The last result enables us to identify a Hilbert space with its dual; this identification is always taken for granted. It is important to remember that the identification is an *anti*-isomorphism since this will make the adjoint of a Hilbert-space operator *seem* a little different from the adjoint of a Banach-space operator.

Exercises

1. Let f be a hermitian form on a real vector space U. Let $V = U + iU$ and extend f to a function $\tilde{f} : V \times V \to \mathbb{C}$ by setting

   $$\tilde{f}(x+iy, z+iw) = f(x,z) + f(y,w) + if(y,z) - if(x,w).$$

   Show that $\tilde{f}$ is a hermitian form on the complex vector space V and $\tilde{f}|U = f$. Show also that $\tilde{f}$ is positive iff f is, and that $\tilde{f}$ is degenerate iff f is.

2. Give an operator $T \in \mathcal{B}(l_2^2)$ such that $(Tx, x) = 0$ for all $x \in l_2^2$ and $\|T\| = 1$.

3. Show that a normed space E is a Euclidean space if and only if the parallelogram law holds in E, i.e.

   $$\|x+y\|^2 + \|x-y\|^2 = 2\|x\|^2 + 2\|y\|^2.$$

4. Let E be a Euclidean space. Show that for $x_1, \ldots, x_n \in E$ we have

   $$\sum_{(\epsilon_i)} \left\| \sum_{i=1}^{n} \epsilon_i x_i \right\|^2 = 2^n \sum_{i=1}^{n} \|x_i\|^2,$$

   where the summation is over all 2^n sequences $(\epsilon_i)_1^n$ ($\epsilon_i = 1$ or -1).

5. Let $u = (\frac{1}{2}, \frac{1}{4}, \frac{1}{8}, \ldots) \in l_2$ and

   $$F = \{x = (x_k)_1^\infty : (x, u) = 0 \text{ and } x_k = 0 \text{ if } k \text{ is large enough}\} \subset l_2.$$

   Show that

   $$F^\perp = \lin\{u\} = \{\lambda u : \lambda \text{ is a scalar}\}.$$

6. Construct a Euclidean space E and a closed subspace $F \subset E$ ($F \neq E$) such that $F^\perp = (0)$. [Note that then, in particular, $E \neq F + F^\perp$.]

7. Let E be a Euclidean space and let $P \in \mathcal{B}(E)$ be a norm-1 projection: $P^2 = P$ and $\|P\| = 1$. Show that P is the orthogonal projection onto $F = \operatorname{Im} P$, i.e. $\operatorname{Ker} P = F^\perp$ and $E = F + F^\perp$.

8. Let A be a non-empty closed convex subset of a Hilbert space H. Show that the distance from A is always attained: for every $b \in H$

there is a unique point $a = a(b) \in A$ such that

$$\|b - a\| = d(b, A) = \inf\{\|b - x\| : x \in A\}.$$

Show also that the map $b \mapsto a(b)$ is continuous.

9. Let x, y and z be points in a Euclidean space. Prove that

$$\|x\| \|y - z\| \leqslant \|y\| \|z - x\| + \|z\| \|x - y\|.$$

10. Let X be the space of continuously differentiable functions on $[0, 1]$, with norm

$$\|f\| = \left(\int_0^1 \{xf^2(x) + 2|f'(x)|^2\} \, dx \right)^{1/2}.$$

Show that X is a Euclidean space. What is the inner product inducing the norm?

11. Let F be a closed subspace of a Hilbert space H. Show that the quotient space H/F is also a Hilbert space.

12. Let Γ be an arbitrary set and let $l_2(\Gamma)$ be the vector space of all functions $f : \Gamma \to \mathbb{C}$ such that $\sum_{\gamma \in \Gamma} |f(\gamma)|^2 < \infty$. In particular, if $f \in l_2(\Gamma)$ then $\{\gamma \in \Gamma : f(\gamma) \neq 0\}$ is countable. Show that $l_2(\Gamma)$ is a Hilbert space with then norm

$$\|f\| = \left(\sum_{\gamma \in \Gamma} |f(\gamma)|^2 \right)^{1/2}.$$

13. Let $\{H_\gamma : \gamma \in \Gamma\}$ be a family of Hilbert spaces. Let H be the vector space of functions $f : \Gamma \to \bigcup_{\gamma \in \Gamma} H_\gamma$ such that $f(\gamma) \in H_\gamma$ and

$$\sum_{\gamma \in \Gamma} \|f(\gamma)\|^2 < \infty.$$

Show that

$$\|f\| = \left(\sum_{\gamma \in \Gamma} \|f(\gamma)\|^2 \right)^{1/2}$$

is a norm on H and with this norm H is a Euclidean space. Is H necessarily a Hilbert space?

14. Turn $C[0, 1]$ into a Euclidean space by setting

$$(f, g) = \int_0^1 f(t) \bar{g}(t) \, dt.$$

Show that this space is incomplete.

15. $f: [0,1] \to \mathbb{R}$ such that $f' \in L_2(0,1)$ and $f(0) = 0$. Prove that V is a real Hilbert space with inner product

$$\langle f, g \rangle = \int_0^1 f'(t)g'(t) \, dt.$$

16. A real $n \times n$ matrix A is called *orthogonal* if $(Ax, Ay) = (x, y)$ for all $x, y \in l_2^n$. Prove that A is orthogonal iff it maps orthogonal vectors into orthogonal vectors and has norm 1.

17. Let P be a bounded linear projection in a Euclidean space E (i.e. $P \in B(E)$ and $P^2 = P$). Show that $\|P\| = 1$ iff $P \neq 0$ and $\operatorname{Im} P \perp \operatorname{Ker} P$.

18. Let A be a non-empty subset of a Hilbert space H and let $T \in \mathcal{L}(H)$ be such that $TH \subset A$ and $(x - Tx) \perp A$ for every $x \in H$. Show that T is a bounded linear operator, A is a closed linear subspace and T is the orthogonal projection onto A: $T = P_A$.

19. Show that the unit ball of l_2 contains an infinite set A such that $\|x - y\| > \sqrt{2}$ for all $x, y \in A$ ($x \neq y$). Show also that A must consist of norm-1 vectors.

Notes

The concept of an abstract Hilbert space was introduced by J. von Neumann, *Eigenwert Theorie Hermitescher funktional Operatoren*, Math. Ann., **102** (1930), 49–131. Earlier special realizations of a Hilbert space had been considered by several people. In particular, from 1904 to 1910 David Hilbert published some fundamental papers on integral equations which led him to consider some 'Hilbert spaces' of functions. It seems that the name 'Hilbert space' was first used by F. Riesz in his book, *Les systèmes d'equations à une infinité d'inconnus*, Paris, 1913, for what we know as l_2: "Considérons l'espace Hilbertien; nous y entendons l'ensemble des systémes (x_k) tels que $\sum |x_k|^2$ converge."

10. ORTHONORMAL SYSTEMS

In this chapter we continue the study of Euclidean spaces and Hilbert spaces. As we shall see, every separable Euclidean space contains the exact analogue of the canonical basis of $l_2^n = (\mathbb{C}^n, \|\cdot\|)$. This means that we can use 'coordinates', which in this context we call Fourier coefficients, to identify the points of the space, these coefficients behaving very much like the coordinates in $\mathbb{C}^n$ or $\mathbb{R}^n$. If our space is in fact a Hilbert space then the space is thus naturally identified with l_2, telling us that all separable Hilbert spaces are isometric.

Let E be a Euclidean space and let $S \subset E$ be a set of vectors. If E is the closed linear span of S, i.e. $\overline{\lim} S = E$, then we call S *fundamental* or *total*. We call S *orthogonal* if it consists of non-zero pairwise orthogonal elements. An orthogonal set of unit vectors is said to be an *orthonormal set*. An orthogonal set is *complete* if it is a maximal orthogonal set. A complete orthonormal set in a Hilbert space is said to be an *orthonormal basis*. We shall see, amongst other results, that every orthonormal basis is a Schauder basis of the best kind, with basis constant 1 (see Exercises 16–18 of Chapter 5), and every Hilbert space contains an orthonormal basis.

Occasionally, we call a set S of vectors a *system* of vectors, especially when S is not written in the form of a sequence.

Theorem 1. A fundamental orthogonal set in a Euclidean space is complete. In a Hilbert space every complete orthogonal set is fundamental.

Proof. (i) Let S be a fundamental set in a Euclidean space E. Then $S^\perp = (\overline{\lim} S)^\perp = E^\perp = (0)$; so if S is orthogonal, then it is complete.

(ii) Let S be a complete orthogonal set in a Hilbert space H. Set $M = \overline{\lim} S$. Then $M^\perp = (0)$ so $M = (M^\perp)^\perp = H$. Thus S is a fundamental system. $\square$

The *Gram–Schmidt orthogonalization process* enables us to replace a sequence of linearly independent vectors by an orthonormal sequence.

Theorem 2. Let $x_1, x_2, \ldots$ be linearly independent vectors in a Euclidean space E. Then there exists an orthonormal sequence $y_1, y_2, \ldots$ such that $\lin\{x_1, \ldots, x_n\} = \lin\{y_1, \ldots, y_n\}$ for every n.

Proof. For $n = 0, 1, \ldots$ set

$$M_n = \lin\{x_1, \ldots, x_n\} \quad \text{and} \quad z_{n+1} = x_{n+1} - P_{M_n} x_{n+1},$$

where P_{M_n} is the orthogonal projection onto M_n, so that $(x - P_{M_n} x) \perp M_n$ for every $x \in E$. Then $z_{n+1} \neq 0$ since $x_{n+1} \notin M_n$ and $P_{M_n} x_{n+1} \in M_n$. Also, $z_{n+1} \perp M_n$. Set $y_{n+1} = z_{n+1} / \|z_{n+1}\|$. Then the sequence $y_1, y_2, \ldots$ has the required properties. Indeed, by the construction, $M_n = \lin\{y_1, \ldots, y_n\}$ and so the y_i are linearly independent. Furthermore, $z_{n+1} \perp M_n$ and so $y_{n+1} \perp M_n = \lin\{y_1, \ldots, y_n\}$. $\qquad\square$

In fact, it is easy to define explicitly the orthogonal projection P_{M_n} appearing in the proof of Theorem 2; namely, set $P_{M_0} = 0$ and for $n \geqslant 1$ define

$$P_{M_n} x = \sum_{i=1}^{n} (x, y_i) y_i.$$

The Gram–Schmidt orthogonalization process enables us to show that every separable space (i.e. every space containing a countable dense set) contains a fundamental orthonormal sequence. Also, by Zorn's lemma every Euclidean space contains a complete orthonormal system.

Theorem 3.

(a) Every separable Euclidean space contains a fundamental orthonormal sequence.

(b) In a Euclidean space, every orthonormal system is contained in a complete orthonormal system. In a Hilbert space, every orthonormal system is contained in an orthonormal basis.

Proof. (a) Let $(x_n)_1^{\infty}$ be a dense sequence in a Euclidean space E. Discard x_n if it is in $\lin\{x_1, \ldots, x_{n-1}\}$ and apply Theorem 2 to the sequence obtained.

(b) Let S_0 be an orthonormal system in a Euclidean space E. Set

$$\Sigma = \{S : S_0 \subset S \text{ and } S \text{ is an orthonormal system in } E\}.$$

Then Σ is partially ordered by inclusion and every totally ordered subset Σ' of Σ has an upper bound in Σ, namely $\bigcup_{S \in \Sigma'} S$. Hence, by Zorn's lemma, Σ has a maximal element S_1. Thus S_1 is a complete orthonormal system containing S_0. If E is a Hilbert space, then the system S_1 is a fundamental system. $\qquad\qquad\qquad\qquad\qquad\qquad\qquad\qquad\square$

Examples 4. (i) Let $\mathbb{T}$ be the unit circle in the complex plane: $\mathbb{T} = \{z \in \mathbb{C}: |z| = 1\}$. Equivalently, $\mathbb{T}$ is the real line modulo 2π:

$$\mathbb{T} = \{e^{it}: t \in \mathbb{R}\} = \{e^{it}: 0 \leqslant t \leqslant 2\pi\}.$$

Let E be the Euclidean space $C(\mathbb{T})$ with inner product

$$(f,g) = \frac{1}{2\pi} \int_0^{2\pi} f(t)\overline{g(t)}\, dt,$$

where we have used the second convention: f and g are continuous functions from $[0,2\pi]$ to $\mathbb{C}$, with $f(0) = f(2\pi)$ and $g(0) = g(2\pi)$. The functions $\varphi_n(t) = e^{int}$ ($n = 0, \pm1, \pm2, \ldots$) form an orthonormal system in E. By the Stone–Weierstrass theorem, the linear span of these functions is dense in $C(\mathbb{T})$ endowed with the uniform norm. Hence $\{\varphi_n : n = 0, \pm1, \ldots\}$ is a fundamental, and so complete, system in E. The completion of E is $L_2(\mathbb{T})$.

(ii) Let E be the Euclidean space $C[-1,1]$ with inner product

$$(f,g) = \int_{-1}^{1} f(t)\overline{g(t)}\, dt.$$

The functions $1, t, t^2, \ldots$ are linearly independent in E and, by the Stone–Weierstrass theorem, form a fundamental system in E. Let $(Q_n(t))_0^\infty$ be the sequence obtained from $1, t, t^2, \ldots$ by the Gram–Schmidt orthogonalization process. Then $Q_n(t)$ is a multiple of the nth *Legendre polynomial*

$$P_n(t) = \frac{1}{2^n n!} D^n([t^2 - 1]^n),$$

where D is the differentiation operator (see Exercise 1).

(iii) $1, \sin t, \cos t, \sin 2t, \cos 2t, \ldots$ is an orthogonal sequence in the Hilbert space $L_2(0, \pi)$. Since the subspace of continuous functions is dense in $L_2(0, \pi)$, by the Stone–Weierstrass theorem this sequence is a fundamental orthogonal sequence.

(iv) For $n = 0, 1, \ldots$ define the nth *Rademacher function* $r_n \in L_2(0, 1)$ by

$$r_n(t) = \begin{cases} 1 & \text{if } \lfloor 2^n t \rfloor \text{ is even} \\ -1 & \text{if } \lfloor 2^n t \rfloor \text{ is odd.} \end{cases}$$

Equivalently, $r_n(t) = \operatorname{sign} \sin 2^n \pi t$. (Note that we view r_n as an element of $L_2(0,1)$; so it is, strictly speaking, an *equivalence class* of functions differing on sets of measure 0. Thus it is also customary to define $r_n(t)$ to be $(-1)^k$ if $k2^{-n} < t < (k+1)2^{-n}$ and leave it undefined for $t = k2^{-n}$.) Then $(r_n(t))_0^\infty$ is an *orthonormal system* in $L_2[0,1]$; it is easily seen to be incomplete.

(v) Let $-\infty \leq a < b \leq \infty$ and let $\varphi \in C(a,b)$ be a positive function. For f and g in the space $C_c(a,b)$ of continuous functions with compact support, set

$$(f,g) = \int_a^b f(t)\overline{g(t)}\varphi(t)\, dt.$$

Then $(\cdot,\cdot)$ is an inner product on $C_c(a,b)$; denote its completion by $L_2(\varphi)$. The space $L_2(\varphi)$ consists of all measurable functions $f(t)$ on $[a,b]$ such that

$$\int_a^b |f(t)|^2 \varphi(t)\, dt < \infty.$$

Now let $a = 0$, $b = \infty$ and $\varphi(t) = e^{-t}$. Then the functions $1, t, t^2, \ldots$ form a fundamental system in $L_2(\varphi)$ and the polynomials obtained from them by the orthogonalization process are the *Laguerre polynomials* up to constant factors, i.e. multiples of $e^t D^n(e^{-t} t^n)$ ($n = 0, 1, \ldots$) (see Exercise 5).

(vi) Consider the space $L_2(\varphi)$ constructed in Example (v) but with $a = -\infty$, $b = \infty$ and $\varphi(t) = e^{-t^2}$. Orthogonalizing the sequence $1, t, t^2, \ldots$ we obtain multiples of the *Hermite polynomials*, i.e. multiples of $e^{t^2} D^n(e^{-t^2})$ ($n = 0, 1, \ldots$) (see Exercise 6). $\quad\square$

If $(\varphi_k)_1^n$ is an orthonormal basis in an n-dimensional Euclidean space E, then every vector $x \in E$ is a linear combination of the φ_k: $x = \sum_{k=1}^n c_k \varphi_k$. Furthermore $c_k = (x, \varphi_k)$. Also, if

$$x = \sum_{k=1}^n c_k \varphi_k \quad \text{and} \quad y = \sum_{k=1}^n d_k \varphi_k,$$

then

$$(x,y) = \sum_{k=1}^n c_k \bar{d}_k \quad \text{and} \quad \|x\|^2 = \sum_{k=1}^n |c_k|^2.$$

In other words, the correspondence $x \leftrightarrow (c_k)_1^n$ identifies E with l_2^n.

The main reason why an orthonormal basis $(\varphi_k)_1^\infty$ in a Hilbert space H is very useful is that analogous assertions hold concerning the representations of vectors in H. As we shall see, for every vector $x \in H$ there are unique coefficients $(c_k)_1^\infty$ such that $x = \sum_{k=1}^\infty c_k \varphi_k$. This sequence $(c_k)_1^\infty$ satisfies $\|x\|^2 = \sum_{k=1}^\infty |c_k|^2$ and every sequence satisfying $\sum_{k=1}^\infty |c_k|^2 < \infty$ arises as a sequence of coefficients. Thus an orthonormal basis enables us to identify H with l_2, just as any n-dimensional Euclidean space can be identified with l_2^n.

Theorem 5. Let $(\varphi_k)_1^\infty$ be an orthonormal sequence in a Hilbert space H. Then, for a scalar sequence $c = (c_k)_1^\infty$, the series $\sum_{k=1}^\infty c_k \varphi_k$ is convergent iff $c \in l_2$ (i.e. $\sum_{k=1}^\infty |c_k|^2 < \infty$). If the series is convergent then

$$\left\| \sum_{k=1}^\infty c_k \varphi_k \right\| = \|c\|_2 = \left(\sum_{k=1}^\infty |c_k|^2 \right)^{1/2}. \tag{1}$$

Proof. Set $x_n = \sum_{k=1}^n c_k \varphi_k$. Then, for $1 \leq n \leq m$, by the Pythagorean theorem, we have

$$\|x_n - x_m\|^2 = \sum_{k=n+1}^m |c_k|^2.$$

Hence $(x_n)_1^\infty$ is convergent iff $\sum_{k=1}^\infty |c_k|^2$ is convergent. Relation (1) holds since, again by the Pythagorean theorem,

$$\|x_n\|^2 = \sum_{k=1}^n |c_k|^2.$$

$\square$

A slightly different formulation of Theorem 5 goes as follows. Let $(\varphi_k)_1^\infty$ be an orthonormal sequence in a Hilbert space H, and let $c = (c_k)_1^\infty \in l_2$. Then there is a vector $x \in H$ such that $(x, \varphi_k) = c_k$ for every k. This is usually called the *Riesz–Fischer theorem*. It looks particularly easy (it *is* particularly easy) because the space is assumed to be complete. The original form of the theorem concerned $L_2[0, 1]$, where the completeness, which is far from trivial, had to be proved.

Let $(\varphi_k)_1^\infty$ be an orthonormal sequence in a Hilbert space H. For $x \in H$, set $c_k = (x, \varphi_k)$ $(k = 1, 2, \ldots)$. We call $c_1, c_2, \ldots$ the *Fourier coefficients of x with respect to* $(\varphi_k)_1^\infty$; the series

$$\sum_{k=1}^\infty c_k \varphi_k$$

is the *Fourier series of x with respect to* $(\varphi_k)_1^\infty$.

It is easily seen that if $(\varphi_k)_1^\infty$ is an orthonormal basis then every vector is the sum of its Fourier series and so, in particular, it is determined by its Fourier coefficients. In fact, we have the following slightly more general result.

Theorem 6. Let $(\varphi_k)_1^\infty$ be an orthonormal sequence in a Hilbert space H and let $M = \overline{\lin}(\varphi_k)_1^\infty$. Then, for all $x, y \in H$, we have

(i) $\displaystyle\sum_{k=1}^\infty (x, \varphi_k)\varphi_k = P_M(x)$;

(ii) $\displaystyle\sum_{k=1}^\infty |(x, \varphi_k)|^2 = \|P_M(x)\|^2 \leq \|x\|^2$;

(iii) $\displaystyle\sum_{k=1}^\infty (x, \varphi_k)\overline{(y, \varphi_k)} = (P_M(x), P_M(y)) = (x, P_M(y)) = (P_M(x), y)$.

Also, suppose $c = (c_k)_1^\infty \in l_2$, i.e. $\sum_{k=1}^\infty |c_k|^2 < \infty$. Then there is a unique vector $u \in M$ with Fourier coefficients $c_k = (u, \varphi_k)$ $(k = 1, 2, \ldots)$, namely $u = \sum_{k=1}^\infty c_k\varphi_k$. Furthermore, $v \in H$ has Fourier coefficients $(c_k)_1^\infty$ iff $u = P_M v$.

Proof. Set

$$x_n = \sum_{k=1}^n (x, \varphi_k)\varphi_k.$$

Then, for $1 \leq k \leq n$, we have $(x - x_n, \varphi_k) = 0$ and so $x = x_n + (x - x_n)$ is a representation of x as a sum of orthogonal vectors. Therefore, by the theorem of Pythagoras,

$$\|x\|^2 = \|x_n\|^2 + \|x - x_n\|^2 \geq \|x_n\|^2 = \sum_{k=1}^n |(x, \varphi_k)|^2.$$

Letting $n \to \infty$, we see that $\sum_{k=1}^\infty |(x, \varphi_k)|^2 < \infty$ so, by Theorem 5, $\sum_{k=1}^\infty (x, \varphi_k)\varphi_k$ is convergent, say to $x' \in M$.

As for every k we have

$$(x - x', \varphi_k) = (x - \lim_{n \to \infty} x_n, \varphi_k) = (x, \varphi_k) - \lim_{n \to \infty}(x_n, \varphi_k)$$

$$= (x, \varphi_k) - (x, \varphi_k) = 0,$$

$x - x'$ is orthogonal to M and so $x' = P_M x$. This proves (i); furthermore, (ii) follows from (i) and (1).

To see (iii), set

$$y_n = \sum_{k=1}^{n} (y, \varphi_k)\varphi_k.$$

Then $P_M y = \lim_{n\to\infty} y_n$ and so

$$(P_M x, y) = (x, P_M y) = (P_M x, P_M y)$$

$$= (\lim_{n\to\infty} x_n, \lim_{n\to\infty} y_n) = \lim_{n\to\infty} (x_n, y_n)$$

$$= \lim_{n\to\infty} \sum_{k=1}^{n} (x, \varphi_k)\overline{(y, \varphi_k)}$$

$$= \sum_{k=1}^{\infty} (x, \varphi_k)\overline{(y, \varphi_k)}.$$

The proof of the second part is equally easy. By Theorem 5, the series $\sum_{k=1}^{\infty} c_k \varphi_k$ is convergent to some vector $u \in M$. Then

$$(u, \varphi_k) = \lim_{n\to\infty} \left(\sum_{l=1}^{n} c_l \varphi_l, \varphi_k \right) = c_k$$

for every k, and so u does have Fourier coefficients $c_1, c_2, \ldots$. Also, if $u = P_M v$ then

$$(v, \varphi_k) = (v, P_M \varphi_k) = (P_M v, \varphi_k) = (u, \varphi_k)$$

and so u and v have the same Fourier coefficients. Finally, if $v \in H$ and $(v, \varphi_k) = c_k$ for every k then $(v - u, \varphi_k) = 0$ for every k and so $v - u$ is orthogonal to $\overline{\lin}(\varphi_k)_1^{\infty} = M$. Therefore $u = P_M v$, as claimed. $\square$

The following useful corollary is amply contained in the result above.

Corollary 7.
  (a) (Bessel's inequality) If $(\varphi_k)_1^{\infty}$ is an orthonormal sequence in a Euclidean space E and $x \in E$ then

$$\sum_{k=1}^{\infty} |(x, \varphi_k)|^2 \leq \|x\|^2.$$

  (b) (Parseval's identities) If $(\varphi_k)_1^{\infty}$ is a complete orthonormal sequence in a Hilbert space H and $x, y \in H$ then

$$\sum_{k=1}^{\infty} |(x, \varphi_k)|^2 = \|x\|^2 \quad \text{and} \quad \sum_{k=1}^{\infty} (x, \varphi_k)(y, \varphi_k) = (x, y). \quad \square$$

Parseval's identities imply that every infinite-dimensional Hilbert space with an orthonormal basis (i.e. a complete orthonormal sequence) is isometric to l_2.

Theorem 8. Let $(\varphi_k)_1^\infty$ be an orthonormal basis in a Hilbert space H. For $x \in H$ define

$$\hat{x}(k) = (x, \varphi_k) \quad \text{and} \quad \hat{x} = (\hat{x}(k))_1^\infty = ((x, \varphi_k))_1^\infty.$$

Then the map $H \to l_2$ given by $x \mapsto \hat{x}$ is a linear isometry of H onto l_2. $\square$

Corollary 9. Every n-dimensional Euclidean space is linearly isometric to l_2^n and every separable infinite-dimensional Hilbert space is linearly isometric to l_2. $\square$

Having seen that all separable infinite-dimensional Hilbert spaces are isometric to l_2, what can we say about other Hilbert spaces? As in Exercise 11 of Chapter 9, given a set Γ, we denote by $l_2(\Gamma)$ the vector space of complex-valued functions f on Γ with countable support and such that $\sum_{\gamma \in \Gamma} |f(\gamma)|^2 < \infty$. Then

$$\|f\| = \left(\sum_{\gamma \in \Gamma} |f(\gamma)|^2 \right)^{1/2}$$

is a norm on $l_2(\Gamma)$ and with this norm $l_2(\Gamma)$ is a Hilbert space. Thus $l_2(\mathbb{N})$ is precisely l_2 and $l_2(\{1, 2, \ldots, n\})$ is l_2^n.

Theorem 10. Every Hilbert space is isometrically isomorphic to a space $l_2(\Gamma)$.

Proof. Let H be a Hilbert space. Then H contains a complete orthonormal system, say $\{\varphi_\gamma : \gamma \in \Gamma\}$. Then, by Bessel's inequality, for every $x \in H$ and for every countable subset Γ_0 of Γ, we have

$$\sum_{\gamma \in \Gamma_0} |(x, \varphi_\gamma)|^2 \leq \|x\|^2.$$

Hence the set $\Gamma_1 = \{\gamma : (x, \varphi_\gamma) \neq 0\}$ is countable,

$$x - \sum_{\gamma \in \Gamma_1} (x, \varphi_\gamma) \varphi_\gamma$$

is orthogonal to every φ_γ and so

$$x = \sum_{\gamma \in \Gamma_1} (x, \varphi_\gamma) \varphi_\gamma.$$

Setting $\hat{x}(\gamma) = (x, \varphi_\gamma)$, the map $H \to l_2(\Gamma)$ given by $x \mapsto \hat{x}$ is a linear isometry of H onto $l_2(\Gamma)$. $\qquad\qquad\qquad\qquad\qquad\qquad\qquad\qquad\qquad$ $\square$

The fact that every separable (infinite-dimensional) Hilbert space is isometrically isomorphic to l_2 is very important when we are studying the abstract properties of Hilbert spaces. Nevertheless, in applications Hilbert spaces often appear as spaces of functions, and then we are interested in the connections between the Hilbert space structure and the properties of the functions.

One of the most important Hilbert function spaces is $L_2(\mathbb{T})$, the completion of $C(\mathbb{T})$, the space of continuous functions on the circle $\mathbb{T}$, defined in Example 4 (i). In the standard orthonormal basis $\{e^{int} : n = 0, \pm 1, \pm 2, \ldots\}$, the kth *Fourier coefficient* c_k of $f \in L_2(\mathbb{T})$ is

$$c_k = \frac{1}{2\pi} \int_0^{2\pi} f(t) e^{-ikt} \, dt$$

and

$$S_n f = \sum_{k=-n}^{n} c_k e^{ikt}$$

is the nth *partial sum*. Then, by Theorem 8,

$$\lim_{n \to \infty} \|S_n f - f\| = 0 \qquad\qquad (2)$$

for every $f \in L_2(\mathbb{T})$, with $\|\cdot\|$ denoting the Hilbert space norm, i.e. the norm in $L_2(\mathbb{T})$. Thus the partial sums *converge in mean square* to f.

For $f \in C(\mathbb{T})$ relation (2) is an immediate consequence of the Stone–Weierstrass theorem, which tell us that f can be uniformly approximated by a trigonometric polynomial. (Of course, we used precisely the Stone–Weierstrass theorem to show that $\{e^{int} : n = 0, \pm 1, \pm 2, \ldots\}$ is a complete system.) In particular, for every $\epsilon > 0$ there is a trigonometric polynomial p such that $|f(t) - p(t)| < \epsilon$ for all t. But if p has degree n (i.e. $p(t) = \sum_{k=-n}^{n} d_k e^{ikt}$) then $S_n p = p$. Since the projection operator S_n has norm 1, we have

$$\|S_n f - f\| = \|(S_n - I)f\| \leq \|(S_n - I)p\| + 2\|f - p\| < 2\epsilon.$$

The very easy relation (2) leaves open the question whether the Fourier series of $f \in L_2(\mathbb{T})$ tends to f pointwise in some sense. For example, it may not be unreasonable to expect that if f is continuous then $(S_n f)(t)$ tends to $f(t)$ for every t. Sadly, this is not true, as was first shown by du Bois Reymond in 1876. Nevertheless, Fejér proved in 1900 that there is a simple way to recover a continuous function from the partial sums of its Fourier series. To be precise, Fejér showed that the Fourier series of a continuous function if *Cesàro summable* and the sum is the function itself: if f is continuous then

$$\frac{1}{n+1} \sum_{k=0}^{n} S_k f,$$

i.e. the *average* of the partial sums, tends uniformly to f. Fejér's theorem was of tremendous importance: it launched the modern theory of Fourier series.

Although $(S_n f)(t)$ need not converge to $f(t)$, the set of points at which it fails to tend to $f(t)$ cannot be too large. It was asked by du Bois Reymond whether the partial sums $S_n f$ of a function $f \in C(\mathbb{T})$ tend to f *almost everywhere*, and later Lusin conjectured that the answer is in the affirmative for every $f \in L_2(\mathbb{T})$. For many years this was one of the most famous conjectures in analysis; finally, it was proved by Carleson in 1966. The very intricate proof is one of the greatest triumphs of hard analysis; not surprisingly, it is far beyond the scope of this book.

Exercises

1. Show that the Legendre polynomials

$$P_n(t) = \frac{1}{2^n n!} D^n([t^2 - 1]^n) \qquad (n = 0, 1, \ldots)$$

   form an orthogonal basis of $L_2(-1, 1)$. Deduce that, up to a positive factor, $P_n(t)$ is the nth term in the sequence obtained from $1, t, t^2, \ldots$ by the Gram–Schmidt orthogonalization process, counting $P_0(t) = 1$ the 0th term, $P_1(t) = t$ the first, $P_2(t) = \frac{1}{2}(3t^2 - 1)$ the second, etc.

2. Show that $P_n(1) = 1$ and $P_n(-1) = (-1)^n$.

3. Deduce from the previous exercises that

$$n P_n(t) - (2n-1) t P_{n-1}(t) + (n-1) P_{n-2}(t) = 0$$

   for all $n \geq 2$.

4. Show that $D((t^2 - 1) P_n'(t))$ is orthogonal to t^k for $k < n$, and deduce that

$$(1 - t^2) P_n''(t) - 2t P_n'(t) + n(n+1) P_n(t) = 0$$

   for every $n \geq 1$.

5. Check that the Laguerre polynomials $e^t D^n(e^{-t}t^n)$ $(n = 0, 1, \ldots)$ are indeed orthogonal polynomials in $L_2(\varphi)$, where $\varphi : [0, \infty) \to \mathbb{R}$ is e^{-t}, as in Example 4 (v).

6. Check the analogous assertion about the Hermite polynomials $e^{t^2} D^n(e^{-t^2})$ and weight function $\varphi = e^{-t^2} : (-\infty, \infty) \to \mathbb{R}$, as in Example 4 (vi).

7. Let $L_2(\varphi)$ be as defined in Example 4 (v), with a positive weight function $\varphi : (a, b) \to \mathbb{R}$. Suppose that $1, t, t^2, \ldots \in L_2(\varphi)$. Let $p_0, p_1, p_2, \ldots$ be the polynomials obtained by orthogonalizing the sequence $1, t, t^2, \ldots$. (We do *not* normalize the sequence; so $p_n(t)$ is the unique monic polynomial of degree n orthogonal to every t^k for $k < n$.)

    (i) Show that

$$(p_n, tp_{n-1}) = \|p_n\|^2.$$

    (ii) Show that for $n \geq 2$ we have

$$p_n(t) = (t - a_n)p_{n-1}(t) - b_n p_{n-2}(t)$$

where

$$a_n = \frac{(tp_{n-1}, p_{n-1})}{\|p_{n-1}\|^2} \quad \text{and} \quad b_n = \frac{\|p_{n-1}\|^2}{\|p_{n-2}\|}.$$

8. Let $x_1, \ldots, x_n$ be unit vectors in a Euclidean space and let $y_1, \ldots, y_n$ be the sequence obtained from it by Gram–Schmidt orthogonalization, with

$$y_i = \sum_{j=1}^{i} \lambda_{ij} x_j.$$

Show that $|\lambda_{ii}| \geq 1$, with equality only if $y_i = \lambda_{ii} x_i$.

9. For $a_i = (a_{ij})_{j=1}^n \in l_2^n$ $(i = 1, \ldots, n)$ let $A = A(a_1, \ldots, a_n) = (a_{ij})_{i,j=1}^n$ be the $n \times n$ complex matrix whose ith row is a_i. Prove *Hadamard's inequality*, stating that

$$|\det A| \leq \prod_{i=1}^{n} \|a_i\|,$$

with equality if and only if either some a_i is 0 or the a_i are orthogonal.

10. Let $x_1, \ldots, x_N$ be linearly independent vectors in a Euclidean space, with $N = \binom{n+1}{2}$. Show that there are orthonormal vectors $y_1, \ldots, y_n$ such that $y_i = \sum_{j \in A_i} \lambda_j x_j$, where $A_1, A_2, \ldots, A_n$ are disjoint subsets of $\{1, 2, \ldots, N\}$.

11. Define the *Gram determinant* of a sequence $x_1, x_2, \ldots, x_n$ in a Hilbert space as

$$G(x_1, \ldots, x_n) = \det\left((x_i, x_j)\right)^n_{i, j=1},$$

i.e. as the determinant of the $n \times n$ matrix whose entry in the ith row and jth column is the inner product (x_i, x_j). Show that $G(x_1, \ldots, x_n) \geq 0$, with equality iff the set $\{x_1, \ldots x_n\}$ is linearly dependent.

Show also that if $\{x_1, \ldots, x_n\}$ is a linearly independent set spanning a subspace M_n then

$$d(x, M_n) = \left(\frac{G(x, x_1, x_2, \ldots, x_n)}{G(x_1, x_2, \ldots, x_n)}\right)^{1/2}.$$

12. Prove the following weak form of the *Riemann–Lebesgue lemma*: if $f \in C(\mathbb{T})$ then

$$\frac{1}{2\pi} \int_0^{2\pi} f(t) e^{-int} \, dt \to 0 \qquad \text{as} \, |n| \to \infty.$$

13. The aim of this exercise is to prove *Fejér's theorem*. For $f \in L_2(\mathbb{T})$ set

$$\sigma_n(f, t) = \frac{1}{n+1} \sum_{k=0}^n (S_k f)(t).$$

(i) Show that

$$\sigma_n(f, t) = \frac{1}{2\pi} \int_0^{2\pi} f(x) K_n(t - x) \, dx,$$

where $K_n(s)$, the *Fejér kernel*, is defined as

$$K_n(s) = \sum_{r=-n}^n \frac{n+1-|r|}{n+1} e^{irs}.$$

(ii) Prove that if $0 < s < 2\pi$ then

$$K_n(s) = \frac{1}{n+1}\left(\frac{\sin \frac{1}{2}(n+1)s}{\sin \frac{1}{2}s}\right)^2$$

and

$$K_n(0) = K_n(2\pi) = n+1.$$

(iii) Deduce from (i) and (ii) that $K_n(s) \geq 0$ for all $s \in \mathbb{T}$, $K_n(s) \to 0$ uniformly for $0 < \delta \leq s \leq 2\pi - \delta < 2\pi$ and

$$\frac{1}{2\pi} \int_0^{2\pi} K_n(s) \, ds = 1.$$

(iv) Deduce Fejér's theorem: if $f \in C(\mathbb{T})$ then

$$\sigma_n(f, t) \to f(t)$$

uniformly in t as $n \to \infty$.

(v) Show also that if $f \in L_\infty(\mathbb{T})$ and f is continuous at t_0 then

$$\sigma_n(f, t_0) \to f(t_0).$$

14+. Let H be a Hilbert space and let $S_1 = \{x_\gamma : \gamma \in \Gamma\}$ and $S_2 = \{y_\gamma : \gamma \in \Gamma\}$ be such that $(x_\alpha, y_\beta) = \delta_{\alpha\beta}$ (where $\delta_{\alpha\beta} = 0$ if $\alpha \neq \beta$ and $\delta_{\alpha\alpha} = 1$). Suppose S_1 is a fundamental system, i.e. $\overline{\lim} S_1 = H$. Does it follow that S_2 is also fundamental?

Notes

The chapter is about the beginning of abstract Fourier analysis. As we have hardly scraped the surface of Fourier analysis proper, the reader is encouraged to consult a book on the topic; T. W. Körner, *Fourier Analysis*, Cambridge University Press, 1988, xii + 591 pp., is particularly recommended.

The original Riesz–Fischer theorem was proved in F. Riesz, *Sur les systèmes orthogonaux de fonctions*, Comptes Rendus, **144** (1907), 615–19 and 734–36, and E. Fischer, *Sur la convergence en moyenne*, Comptes Rendus, **144** (1907), 1022–4. It was a clear case of independent discovery, precisely described by Fischer in the introduction of his paper: "Le 11 mars, M. Riesz a présenté à l'Académie une Note sur les systèmes orthogonaux de fonctions (Comptes Rendus, 18 mars 1907). J'étais arrivé au même résultat et je l'ai démontré dans une conférence fait à la Société mathématique à Brünn, déja le 5 mars. Ainsi mon indépendance est évidente, mais la priorité de la publication revient à M. Riesz."

Carleson's theorem, mentioned at the end of the chapter, was proved in L. Carleson, *Convergence and growth of partial sums of Fourier series*, Acta Math., **116** (1966), 135–57.

Exercise 9 is from J. Hadamard, *Résolution d'une question relative aux déterminants*, Bulletin Sci. Math. (2), **17** (1893), 240–348, and Exercise

Chapter 10: Orthonormal systems

13 is from L. Fejér, *Sur les fonctions bornées et integrables*, Comptes Rendus, **131** (1900), 984–7 and *Investigations of Fourier series* (in Hungarian), Mat. és Fiz. Lapok, **11** (1902), 49–68; see also *Untersuchungen über Fouriersche Reihen*, Math. Annalen, **58** (1904), 51–69.

11. ADJOINT OPERATORS

In the next four sections we give a brief account of the theory of bounded linear operators on Banach spaces. Our aim is to present several general concepts and prove some of the fundamental results. We are mainly interested in the spectral theory, to be treated in three sections, but before we embark on that, in this chapter we study the basic properties of the adjoint of an operator.

Throughout this chapter we shall use the product notation for the value of a linear functional on an element: $\langle x, f \rangle = \langle f, x \rangle = f(x)$ for x in a vector space V and f in V', the dual of V. In particular, if X is a normed space and X^* is the dual of X, i.e. the Banach space of all bounded linear functionals on X, then $\langle \cdot, \cdot \rangle$ is the bilinear form on $X \times X^*$ given by $\langle x, f \rangle = f(x)$. Thus

$$\langle \lambda x + \mu y, f \rangle = \lambda \langle x, f \rangle + \mu \langle y, f \rangle \quad \text{and} \quad \langle x, \lambda f + \mu g \rangle = \lambda \langle x, f \rangle + \mu \langle x, g \rangle.$$

(Note the absence of conjugates: $\langle \cdot, \cdot \rangle$ is a *bilinear* form and *not a hermitian* form. This is why it is not confusing to have $\langle x, f \rangle = \langle f, x \rangle$.) Recall that $|\langle x, f \rangle| \leqslant \|x\| \|f\|$ for all $x \in X$ and $f \in X^*$. Furthermore,

$$\|x\| = \sup\{|\langle x, f \rangle| : f \in B(X^*)\} = \max\{|\langle x, f \rangle| : f \in B(X^*)\}$$

and

$$\|f\| = \sup\{|\langle x, f \rangle| : x \in B(X)\}.$$

Let X and Y be normed spaces. Recall from Chapter 1 that the *adjoint* or *dual* T^* of an operator $T \in \mathcal{B}(X, Y)$ is the unique map $T^* : Y^* \to X^*$ such that

$$\langle x, T^*g \rangle = \langle Tx, g \rangle \tag{1}$$

for all $x \in X$ and $g \in Y^*$. Indeed, for $g \in Y^*$, we have a function T^*g on X defined by $(T^*g)(x) = \langle Tx, g \rangle$; it is easily seen that this function is linear and, in fact, $T^*g \in Y^*$. In turn, the *second dual* $(T^*)^*$ of T is the unique map $T^{**}: X^{**} \to Y^{**}$ such that

$$\langle T^{**}\varphi, g \rangle = \langle \varphi, T^*g \rangle$$

for all $\varphi \in X^{**}$ and $g \in Y^*$.

Let us summarise the basic properties of taking adjoints.

Theorem 1. Let X and Y be normed spaces and let $T, T_1, T_2 \in \mathscr{B}(X, Y)$. Then
  (a) $T^* \in \mathscr{B}(Y^*, X^*)$ and $\|T^*\| = \|T\|$;
  (b) for scalars λ_1 and λ_2, we have $(\lambda_1 T_1 + \lambda_2 T_2)^* = \lambda_1 T_1^* + \lambda_2 T_\lambda^*$;
  (c) with the natural inclusions $X \subset X^{**}$ and $Y \subset Y^{**}$ we have $T^{**}|X = T$, i.e. $T^{**}x = Tx$ for all $x \in X$;
  (d) if Z is a normed space and $S \in \mathscr{B}(Y, Z)$ then $(ST)^* = T^*S^*$;
  (e) if T is invertible, i.e. $T^{-1} \in \mathscr{B}(Y, X)$ then T^* is also invertible and $(T^*)^{-1} = (T^{-1})^*$.

Proof. Part (a) is Theorem 3.9, part (b) is immediate from the definition of the adjoint, and (c) follows from Theorem 3.10, giving the natural embeddings $X \to X^{**}$ and $Y \to Y^{**}$. To see (d), note that for $x \in X$ and $h \in Z^*$ we have

$$\langle x, T^*S^*h \rangle = \langle Tx, S^*h \rangle = \langle STx, h \rangle.$$

Finally, (e) follows since $T^{-1}T = I_X$, where I_X is the identity operator on X and so $I_{X^*} = (T^{-1}T)^* = T^*(T^{-1})^*$. Similarly, $I_{Y^*} = (T^{-1})^*T^*$. Hence $(T^*)^{-1} = (T^{-1})^*$. $\qquad\qquad\qquad\qquad\qquad\qquad\qquad\square$

Let now H and K be Hilbert spaces, with their inner products written as $(\cdot, \cdot)$. We know from the Riesz representation theorem in Chapter 8 that a Hilbert space is naturally isometric with its dual but the isometry is an anti-isomorphism. Thus $x \in H$ can be considered as a functional on H: it acts on H as multiplication on the right: $(\cdot, x)$. Then for $T \in \mathscr{B}(H, K)$ we have $T^* \in \mathscr{B}(K, H)$.

Equivalently, T^* is defined by

$$(x, T^*y) = (Tx, y) \qquad\qquad\qquad (2)$$

for all $x \in H$ and $y \in K$. Indeed, for a fixed $y \in K$ the function $f: H \to \mathbb{C}$ defined by $f(x) = (Tx, y)$ is a bounded linear functional on H. Hence, by the Riesz representation theorem, $f(x) = (x, u)$ for some unique vector $u \in H$. We *define* T^*y to be this vector u. It is immediate that T^* is a linear map from K to H. Since

$$|f(x)| \leq \|Tx\| \|y\| \leq \|T\| \|y\| \|x\|,$$

we have

$$\|u\| = \|T^*y\| = \|f\| \leq \|T\| \|y\|$$

and so T^* is a bounded linear map and $\|T^*\| \leq \|T\|$.

Theorem 2. Let H and K be Hilbert spaces and let $T, T_1, T_2 \in \mathfrak{B}(H, K)$. Then $T^* \in \mathfrak{B}(K, H)$, $\|T^*\| = \|T\|$ and, for all scalars λ_1 and λ_2 we have

$$(\lambda_1 T_1 + \lambda_2 T_2)^* = \bar{\lambda}_1 T_1^* + \bar{\lambda}_2 T_2^*, \tag{3}$$

$$(T_1 T_2)^* = T_2^* T_1^* \tag{4}$$

and

$$T^{**} = T. \tag{5}$$

Furthermore, if $K = H$ then

$$\|T\|^2 = \|TT^*\| = \|T^*T\| = \|T^*\|^2. \tag{6}$$

Proof. We known that $\|T^*\| = \|T\|$ and $(T_1 T_2)^* = T_2^* T_1^*$; furthermore $T^{**} = T$ since H is reflexive. To see (3), note that for $x \in H$ and $y \in K$ we have

$$\begin{aligned}
(x, (\lambda_1 T_1 + \lambda_2 T_2)^* y) &= ((\lambda_1 T_1 + \lambda_2 T_2)x, y) \\
&= \lambda_1 (T_1 x, y) + \lambda_2 (T_2 x, y) \\
&= \lambda_1 (x, T_1^* y) + \lambda_2 (x, T_2^* y) \\
&= (x, \bar{\lambda}_1 T_1^* y) + (x, \bar{\lambda}_2 T_2^* y) \\
&= (x, (\bar{\lambda}_1 T_1^* + \bar{\lambda}_2 T_2^*) y).
\end{aligned}$$

Of course, this is also clear from Theorem 1(b) and the fact that the natural isometry between a Hilbert space and its dual is an *anti-isomorphism*.

Finally, let $T \in \mathfrak{B}(H)$. Then

$$\begin{aligned}
\|T\|^2 = \sup_{\|x\|=1} \|Tx\|^2 &= \sup_{\|x\|=1} |(Tx, Tx)| \\
&= \sup_{\|x\|=1} |(T^*Tx, x)| \\
&\leq \|T^*T\| \leq \|T^*\| \|T\| = \|T\|^2. \qquad \square
\end{aligned}$$

Let us emphasize again that in (3) we have to take the conjugates of the coefficients, unlike in Theorem 1(b), which is the analogous statement for normed spaces. This is because in (3) we take T_i^* as a map from K to H, rather than from K^* to H^*, as in the normed-space case; the conjugates appear because H and H^* are identified by an *anti*-isomorphism. Theorem 2 is of special interest in the case $H = K$. An *involution* on a complex Banach algebra is a map $x \mapsto x^*$ such that (3), (4) and (5) hold:

$$(\lambda_1 x_1 + \lambda_2 x_2)^* = \bar{\lambda}_1 x_1^* + \bar{\lambda}_2 x_2^*, \qquad (x_1 x_2)^* = x_2^* x_1^*, \qquad x^{**} = x.$$

A *C*-algebra* is a Banach algebra with an involution satisfying (6), i.e. in which $\|xx^*\| = \|x\|^2$ (and so $\|x\|^2 = \|x^*x\| = \|x^*\|^2$). Thus Theorem 2 claims that $\mathcal{B}(H)$ is a *C*-*algebra with involution $T \mapsto T^*$. Hence every norm-closed subalgebra of $\mathcal{B}(H)$ which is closed under taking adjoints (i.e. every closed **-*subalgebra of $\mathcal{B}(H)$) is also a *C*-*algebra. What is remarkable is that the converse of this statement is also true: every *C*-*algebra is isomorphic to a closed **-*subalgebra of $\mathcal{B}(H)$. This is the celebrated *Gelfand–Naĭmark theorem*; although it is beyond the scope of this book, we shall present some exercises concerning it at the end of the next chapter. Given a normed space X with dual X^*, for a subspace $K \subset X$ define the *annihilator of K in X** as

$$K^0 = \{f \in X^* : \langle x, f \rangle = 0 \text{ for all } x \in K\}.$$

Similarly, for a subspace $L \subset X^*$, the *annihilator of L in X* (or the *preannihilator* of L) is

$$^0L = \{x \in X : \langle x, f \rangle = 0 \text{ for all } f \in L\}.$$

Strictly speaking, K^0 usually denotes the *polar* of a set $K \subset X$:

$$K^0 = \{f \in X^* : |\langle x, f \rangle| \leqslant 1 \text{ for all } f \in L\}$$

and ^0L is the *prepolar* of a set $L \subset X^*$ (or the *polar of L in X*):

$$^0L = \{x \in X : |\langle x, f \rangle| \leqslant 1 \text{ for all } f \in \}.$$

Of course, if K and L are subspaces, as we have chosen them, or at least if they are unions of one-dimensional subspaces, then the two definitions coincide (see Exercise 1).

It is clear that for any sets $K \subset X$ and $L \subset X^*$ the annihilators K^0 and ^0L are closed; furthermore,

$$K^0 = (\operatorname{lin} K)^0 = (\overline{\operatorname{lin}} K)^0 \qquad \text{and} \qquad ^0L = {}^0(\operatorname{lin} L) = {}^0(\overline{\operatorname{lin}} L).$$

Theorem 3. Let X and Y be normed spaces and let $T \in \mathcal{B}(X, Y)$. Then

$$\operatorname{Ker} T = {}^0(\operatorname{Im} T^*) \qquad \text{and} \qquad \operatorname{Ker} T^* = (\operatorname{Im} T)^0. \qquad \square$$

Proof. Clearly

$$\operatorname{Ker} T = \{x \in X^* : Tx = 0\}$$
$$= \{x \in X : \langle Tx, g \rangle = 0 \text{ for all } g \in Y^*\}$$
$$= \{x \in X : \langle x, T^*g \rangle = 0 \text{ for all } g \in Y^*\} = {}^0(\operatorname{Im} T^*).$$

Similarly,

$$\operatorname{Ker} T^* = \{g \in Y^* : T^*g = 0\}$$
$$= \{g \in Y^* : \langle x, T^*g \rangle = 0 \text{ for all } x \in X\}$$
$$= \{g \in Y^* : \langle Tx, g \rangle = 0 \text{ for all } x \in X\} = (\operatorname{Im} T)^0. \qquad \square$$

Note that if $L \subset X^*$ then ${}^0L \subset X$ and $L^0 \subset X^{**}$ and so, in general, we cannot expect ${}^0L$ to be equal to L^0 under the natural inclusion $X \subset X^{**}$. However, if X if reflexive and so X is *identified* with X^{**} then for every set $L \subset X^*$ we have ${}^0L = L^0$. If H is a Hilbert space then not only is H a reflexive space but also the dual H^* is identified with H. With this identification, $L^0 = {}^0L = L^\perp$ for every set $L \subset H$. Hence Theorem 3 has the following immediate consequence.

Corollary 4. Let H and K be Hilbert spaces and let $T \in \mathcal{B}(H, K)$. Then

$$\operatorname{Ker} T = (\operatorname{Im} T^*)^\perp \qquad \text{and} \qquad \operatorname{Ker} T^* = (\operatorname{Im} T)^\perp. \qquad \square$$

It is worth noting that $\operatorname{Im} T = {}^0(\operatorname{Ker} T^*)$ does not hold in general since $\operatorname{Im} T$ need not be closed. However, if $\operatorname{Im} T$ is closed then we do have $\operatorname{Im} T = {}^0(\operatorname{Ker} T^*)$ (see Exercise 3).

Let H be a Hilbert space. An operator $T \in \mathcal{B}(H)$ is called *hermitian* or *self-adjoint* if $T^* = T$. Thus T is self-adjoint iff

$$(Tx, y) = (x, Ty) \text{ for all } x, y \in H.$$

Clearly an operator $T \in \mathcal{B}(H)$ is self-adjoint iff $\langle x, y \rangle = (Tx, y)$ is a hermitian form on H. If S and T are commuting self-adjoint operators then ST is also hermitian since

$$(STx, y) = (Tx, Sy) = (x, TSy) = (x, STy).$$

In particular, if T is self-adjoint then so is T^n for every $n \geq 1$. Note also that if T is self-adjoint then by Theorem 2,

$$\|T^2\| = \|T^*T\| = \|T\|^2, \qquad \|T^4\| = \|T^2\|^2 = \|T\|^4,$$

etc. Therefore $\|T^{2^k}\| = \|T\|^{2^k}$. Also, if $1 \leq n \leq 2^k$ then

$$\|T^{2^k}\| = \|T^n T^{2^k-n}\| \leq \|T^n\| \|T\|^{2^k-n} \leq \|T\|^{2^k}$$

and so $\|T^n\| = \|T\|^n$.

If T is self-adjoint then (Tx, x) is real for every $x \in H$ since $(Tx, x) = \overline{(x, Tx)}$. We call a self-adjoint operator T *positive* if $(Tx, x) \geq 0$ for every $x \in H$.

Note that, for $T \in \mathcal{B}(H)$, the operator T^*T is a positive self-adjoint operator. Indeed,

$$(T^*T)^* = T^*(T^*)^* = T^*T \quad \text{and} \quad (T^*Tx, x) = (Tx, Tx) = \|Tx\|^2 \geq 0.$$

Replacing T by T^*, we see that TT^* is also a positive self-adjoint operator.

Theorem 5. Let H be a complex Hilbert space. Then every operator $T \in \mathcal{B}(H)$ has a representation in the form $T = T_1 + iT_2$, where T_1 and T_2 are hermitian, and this representation is unique.

Proof. Set

$$T_1 = \tfrac{1}{2}(T + T^*) \quad \text{and} \quad T_2 = -\tfrac{1}{2}i(T - T^*).$$

Then T_1 and T_2 are hermitian and $T = T_1 + iT_2$. The uniqueness follows from the fact that if T_1 and T_2 are hermitian and $T_1 + iT_2 = 0$ then

$$T_1 + iT_2 = (T_1 + iT_2)^* = T_1 - iT_2$$

and so $T_2 = 0$ and $T_1 = 0$. □

Examples 6. (i) Let T be the *right translation* on l_2, i.e. let $T((x_1, x_2, \ldots)) = ((0, x_1, x_2, \ldots))$. Then T^* is the *left translation*: $T^*((x_1, x_2, \ldots)) = ((x_2, x_3, \ldots))$. Clearly $\|T\| = \|T^*\| = 1$, T^*T is the identity I but $TT^* \neq I$:

$$TT^*((x_1, x_2, x_3, \ldots)) = ((0, x_2, x_3, \ldots)).$$

(ii) For $\varphi \in C[0,1]$ define $T_\varphi : L_2(0,1) \to L_2(0,1)$ as multiplication by φ:

$$(T_\varphi f)(t) = \varphi(t) f(t) \qquad (0 \leq t \leq 1).$$

Then

$$T_\varphi^* = T_{\bar\varphi} \quad \text{and} \quad \|T_\varphi\| = \|T_{\bar\varphi}\| = \max\{|\varphi(t)| : 0 \leqslant t \leqslant 1\}.$$

Clearly T_φ is a positive hermitian operator iff φ is a non-negative real-valued function.

(iii) Let M be a closed subspace of a Hilbert space and let P_M be the orthogonal projection onto M. Then P_M is a positive self-adjoint operator and (as every projection) satisfies $P_M^2 = P_M$. □

It is easily seen that the properties in Example 6 (iii) characterise the orthogonal projections in a Hilbert space.

Theorem 7. Let H be a Hilbert space and let $P \in \mathcal{B}(H)$ be a self-adjoint projection: $P^2 = P = P^*$. Then $M = \text{Im}\, P$ is closed and P is the orthogonal projection of H onto M: $P = P_M$.

Proof. Since P is a projection, we have $x - Px \in \text{Ker}\, P$ and $x = (x - Px) + Px$ for every $x \in H$. So $H = \text{Ker}\, P + \text{Im}\, P$. By Corollary 4, we have $\text{Ker}\, P = (\text{Im}\, P^*)^\perp = (\text{Im}\, P)^\perp$ and so H is the orthogonal direct sum of $\text{Ker}\, P$ and $\text{Im}\, P$. □

In addition to hermitian operators and orthogonal projections, let us introduce two other important classes of operators. Given a Hilbert space H, an operator $T \in \mathcal{B}(H)$ is said to be *normal* if $TT^* = T^*T$, and *unitary* if T is invertible and its inverse is T^*. Note that every hermitian operator is normal, and so is every unitary operator. In the following results we characterize normal and unitary operators.

Theorem 8. Let H be Hilbert space and $T \in \mathcal{B}(H)$.
 (a) T is normal iff $\|Tx\| = \|T^*x\|$ for all $x \in H$.
 (b) If T is normal then $\|T^n\| = \|T\|^n$ for every $n \geqslant 1$ and

$$\text{Ker}\, T = \text{Ker}\, T^* = (\text{Im}\, T)^\perp = (\text{Im}\, T^*)^\perp.$$

Proof. (a) Clearly,

$$\|Tx\|^2 - \|T^*x\|^2 = (Tx, Tx) - (T^*x, T^*x)$$

$$= (T^*Tx, x) - (TT^*x, x) = ((T^*T - TT^*)x, x).$$

From Theorem 9.2 we know that $T^*T - TT^* = 0$ iff $((T^*T - TT^*)x, x) = 0$ for every x.

(b) If T is normal then, by (a), we have $\text{Ker}\, T = \text{Ker}\, T^*$. Hence, by Corollary 4,

$$\text{Im } T^{\perp} = \text{Ker } T^* = \text{Ker } T = (\text{Im } T^*)^{\perp}.$$

Furthermore, as T^*T is hermitian,

$$\|T\|^{2n} = \|T^*T\|^n = \|(T^*T)^n\|,$$

and so

$$\|T\|^{2n} = \|(T^*T)^n\| \leqslant \|(T^*)^n\| \|T^n\| \leqslant \|T\|^n \|T^n\| \leqslant \|T\|^{2n},$$

implying $\|T^n\| = \|T\|^n$. $\qquad\qquad\square$

As a consequence of Theorem 8 one can see that Theorem 7 can be strengthened: if a projection is normal then it is an orthogoanl projection (see Exercise 7).

Theorem 9. Let H be a Hilbert space and let $U \in \mathcal{B}(H)$ be such that $\text{Im } U = H$. Then the following are equivalent:
 (a) U is unitary;
 (b) U is an *isometry*: $\|Ux\| = \|x\|$ for every $x \in H$;
 (c) U preserves the inner product: $(Ux, Uy) = (x, y)$ for all $x, y \in H$.

Proof. The polarization identity (3) in Chapter 9 implies that U is an isometry iff it preserves the inner product. Thus (b) and (c) are equivalent.
 If U is unitary then

$$(Ux, Uy) = (U^*Ux, y) = (x, y).$$

Conversely, if $(Ux, Uy) = (x, y)$ for all $x, y \in H$ then $(U^*Ux, y) = (x, y)$ and so $U^*U = I$. Since $\text{Im } U = H$ and U is an isometry, U is invertible. Therefore $U^* = U^{-1}$. $\qquad\qquad\square$

Our last aim in this chapter is to show that the converse of Theorem 1 (v) also holds if X is a Banach space, i.e. T^* is invertible iff T is. First we note a simple condition for invertibility. Call $T \in \mathcal{B}(X, Y)$ *bounded below* if $\|Tx\| \geqslant \epsilon \|x\|$ for all $x \in X$ and some $\epsilon > 0$.

Theorem 10. Let X be a Banach space, Y a normed space, and let $T \in \mathcal{B}(X, Y)$. Then $T^{-1} \in \mathcal{B}(Y, X)$ iff $\text{Im } T$ is dense in Y and T is bounded below.

Proof. The necessity of the two conditions is obvious. Suppose then that the conditions are satisfied. If T is bounded below then it is injective, so $T^{-1} \in \mathcal{L}(Z, X)$, where $Z = \text{Im } T$. Since Z is dense in Y, for every $y \in Y$ there is a sequence $(z_k)_1^\infty$ in Z converging to y. Then by the second condition, $(T^{-1}z_k)_1^\infty$ is also convergent, say to x. Hence

$$Tx = \lim_{k \to \infty} T(T^{-1}z_k) = \lim_{k \to \infty} z_k = y$$

and so $Y = Z$.

This shows that $T^{-1} \in \mathcal{L}(Y, X)$. If $\|Tx\| \geqslant \epsilon\|x\|$ for all $x \in X$ then, clearly, $\|T^{-1}\| \leqslant 1/\epsilon$. $\square$

Theorem 11. Let X be a Banach space, Y a normed space and let $T \in \mathcal{B}(X, Y)$. Then T^* is invertible iff T is.

Proof. We have seen that if T is invertible then so is T^*. Suppose then that T^* is invertible. Let us check that the two conditions in Theorem 10 are satisfied.

By Theorem 3 we have $(\operatorname{Im} T)^0 = \operatorname{Ker} T^* = (0)$ and so $\operatorname{Im} T$ is dense in Y. To see that T is bounded below, let $x \in X$ and let f be a support functional at x, i.e. let $f \in X^*$ be such that $\langle x, f \rangle = \|x\|$ and $\|f\| = 1$. Then

$$\|x\| = \langle x, f \rangle = \langle x, T^*(T^*)^{-1}f \rangle$$
$$= \langle Tx, (T^*)^{-1}f \rangle$$
$$\leqslant \|Tx\|\|(T^*)^{-1}f\| \leqslant \|Tx\|\|(T^*)^{-1}\|.$$

Therefore $\|Tx\| \geqslant \|(T^*)^{-1}\|^{-1}\|x\|$, completing the proof. $\square$

If $\operatorname{Im} T$ is not dense in Y, say $TX \subset Z$, with Z a closed subspace, then $Z \subset \operatorname{Ker} f$ for some $f \in Y^*$ ($f \neq 0$). Hence $\langle x, T^*f \rangle = \langle Tx, f \rangle = 0$ for all $x \in X$ and so $T^*f = 0$. In particular, if T^* is bounded below then $\operatorname{Im} T$ is dense in Y. This gives us yet another condition for invertibility; let us state it together with Theorems 10 and 11.

Theorem 12. Let X be a Banach space, Y a normed space, and let $T \in \mathcal{B}(X, Y)$. Then the following conditions are equivalent:
 (a) T is invertible;
 (b) T^* is invertible;
 (c) $\operatorname{Im} T$ is dense in Y and T is bounded below;
 (d) T and T^* are both bounded below. $\square$

The question of invertibility brings us to the study of the spectrum of an operator and the structure of the algebra of bounded linear operators. But that requries a new chapter.

Exercises

1. Given a Banach space X and a set $K \subset X$, the *annihilator* of K in X^* is

$$K^a = \{f \in X^* : \langle x, f \rangle = 0 \text{ for all } x \in K\}$$

   and the *annihilator of a set* $L \subset X^*$ *in* X (or the *preannihilator* of L) is

$$^aL = \{x \in X : \langle x, f \rangle = 0 \text{ for all } f \in L\}.$$

   Show that if K and L are subspaces then $K^a = K^0$ and $^aL = {}^0L$, where K^0 is the polar of K and ^0L is the prepolar of L. Show also that

$$K^a = (\operatorname{lin} K)^a = (\overline{\operatorname{lin}} K)^a = (\operatorname{lin} K)^0$$

   and

$$^aL = {}^a(\operatorname{lin} L) = {}^a(\overline{\operatorname{lin}} L) = {}^0(\operatorname{lin} L).$$

2. Give examples showing that for a Banach space X and a subspace $L \subset X^*$, the sets ^0L and L^0 need not be equal under the natural inclusion $X \subset X^{**}$.

3. Let X and Y be normed spaces and $T \in \mathcal{B}(X, Y)$. Show that $^0(\operatorname{Ker} T^*)$ is the closure of $\operatorname{Im} T$.

4. A subspace U of a normed space V is said to be an *invariant subspace* of an operator $S \in \mathcal{B}(V)$ if $SU \subset U$, i.e. $Su \in U$ for all $u \in U$. Let X be a normed space and $T \in \mathcal{B}(X)$. Show that a closed subspace Y of X is an invariant subspace of T iff Y^0 is an invariant subspace of T^*.

5. Let X and Y be Banach spaces and $T \in \mathcal{B}(X, Y)$. Prove that $\operatorname{Im} T$ is closed iff $\operatorname{Im} T^*$ is closed.

6. Let X be a Banach space. Show that for $T \in \mathcal{B}(X)$ the series $\sum_{n=0}^{\infty} T^n/n!$ converges in norm to an element of $\mathcal{B}(X)$, denoted by $\exp T$. Show also that $(\exp T)^* = \exp T^*$ and if $S \in \mathcal{B}(X)$ commutes with T then

$$(\exp S)(\exp T) = (\exp T)(\exp S)\exp(S + T).$$

In the exercises below, H denotes a Hilbert space.

7. Let T be a bounded linear operator on a Hilbert space H. Show that T has an eigenvector iff T^* has 1-codimensional closed invariant subspace.

8. Let H be a Hilbert space and $P \in \mathscr{B}(H)$ a projection: $P^2 = P$. Show that the following are equivalent:
    (a) P is an orthogonal projection;
    (b) P is hermitian;
    (c) P is normal;
    (d) $(Px, x) = \|Px\|^2$ for all $x \in H$.

9. Let $U \in \mathscr{B}(H)$ be a unitary operator.
    (i) Show that $\operatorname{Im}(U - I) = \operatorname{Im}(U^* - I)$ and deduce that $\operatorname{Ker}(U - I) = (\operatorname{Im}(U - I))^{\perp}$.
    (ii) For $n \geqslant 1$ set

$$S_n = \frac{1}{n}(I + U + \cdots + U^{n-1}).$$

Show that $S_n x \to P_M x$ for every $x \in H$, where $M = \operatorname{Ker}(U - I)$. (One expresses this by saying that S_n tends to P_M in the *strong operator topology*.)

10. Show that if $T \in \mathscr{B}(H)$ is hermitian then $\exp iT$ is unitary.

11. The aim of this exercise is to prove the Fuglede–Putnam theorem. Suppose that $R, S, T \in \mathscr{B}(H)$, with R and T normal and $RS = ST$.
    (i) Show that

$$(\exp R)S = S(\exp T).$$

(ii) By considering $\exp(R^* - R) S \exp(T - T^*)$, show that

$$\|(\exp R^*) S \exp(-T^*)\| \leqslant \|S\|.$$

(iii) For $f \in (\mathscr{B}(H))^*$ and $\lambda \in \mathbb{C}$ set

$$F(\lambda) = f(\exp(\lambda R^*) S \exp(-\lambda T^*)).$$

Show that $F(\lambda)$ is an analytic function and $|F(\lambda)| \leqslant \|f\| \|S\|$ for every $\lambda \in \mathbb{C}$. Apply Liouville's theorem to deduce that $F(\lambda) = F(0) = f(S)$ for every λ and hence that

$$\exp(\lambda R^*) S = S \exp(\lambda T^*).$$

(iv) Deduce that $R^* S = ST^*$.

Notes

The notion of an adjoint operator was first introduced by S. Banach, *Sur les fonctionelles linéaires II*, Studia Math., **1** (1929), 223–39. Our treatment of adjoint operators is standard. The Gelfand–Naĭmark theorem was proved in I. M. Gelfand and M. A. Naĭmark, *On the embedding of normed rings into the ring of operators in Hilbert space*, Mat. Sbornik N.S., **12** (1943), 197–213; for a thorough treatment of the subject see S. Sakai, *C*-algebras and W*-algebras*, Springer- Verlag, New York-Heidelberg-Berlin, 1971. For the Fuglede–Putnam theorem in Exercise 11, see chapter 41 in P. R. Halmos, *Introduction to Hilbert space and the theory of spectral multiplicity*, Chelsea, New York, 1951.

12. THE ALGEBRA OF BOUNDED LINEAR OPERATORS

In this chapter we shall consider *complex* Banach spaces and *complex unital* Banach algebras, as we shall study the spectra of various elements. Recall that a *complex unital Banach algebra* is a complex algebra A with an identity e, which is also a Banach space, in which the algebra structure and the norm are connected by $\|e\| = 1$ and $\|ab\| \leqslant \|a\| \|b\|$ for all $a, b \in A$. If there is an *involution* $x \mapsto x^*$ in A such that $x^{**} = x$, $(x+y)^* = x^* + y^*$, $(\lambda x)^* = \bar{\lambda} x^*$, $(xy)^* = y^* x^*$ and $\|x^* x\| = \|x\|^2$ then A is a *C*-algebra*. As we noted earlier, if X is a complex Banach space then $\mathcal{B}(X)$ is a complex unital Banach algebra, and if H is a complex Hilbert space then $\mathcal{B}(H)$ is a *C*-algebra.*

An element a of a Banach algebra A is *invertible* (in A) if $ab = ba = e$ for some $b \in A$; the (unique) element b is the *inverse* of a, and is denoted by a^{-1}. The *spectrum* of $a \in A$ is

$$\sigma_A(a) = Sp_A(a) = \{\lambda \in \mathbb{C} : \lambda e - a \text{ is not invertible in } A\},$$

and the *resolvent set* of a is $\delta_A(a) = \mathbb{C} \backslash \sigma_A(a)$. A point of $\delta_A(a)$ is said to be a *regular point*. The function $R : \delta(a) \to A$ given by $R(\lambda) = (\lambda e - a)^{-1}$ is the *resolvent* of a. The element $R(\lambda)$ is the *resolvent of a at λ* or, with a slight abuse of terminology, the *resolvent* of a.

The prime example of a Banach algebra we are interested in is the algebra $\mathcal{B}(X)$ of bounded linear operators on a complex Banach space X; so our algebra elements are operators. In view of this, if $T \in \mathcal{B}(X)$ then we define the *spectrum* and *resolvent set* of T without any reference to $\mathcal{B}(X)$:

$$\sigma(T) = \{\lambda \in \mathbb{C} : \lambda I - T \text{ is not invertible}\}$$

where I is the identity operator on X, and

$$\rho(T) = \mathbb{C} \backslash \sigma(T).$$

If A is a complex unital Banach algebra then A can be considered to be a subalgebra of $\mathcal{B}(A)$, the algebra of all bounded linear operators acting on the Banach space A, with the element a corresponding to the operator L_a of left multiplication by a (so that $a \mapsto L_a$, where $L_a(x) = ax$ for every $x \in A$). In particular if $a \in A$ is invertible then so is $L_a \in \mathcal{B}(A)$, with inverse $L_{a^{-1}}$. Conversely, if $S \in \mathcal{B}(A)$ is the inverse of L_a, so that

$$x = (L_a S)x = a(Sx)$$

for every $x \in A$, then with $b = Se$ we have $ab = 1$ and so $a(ba-e) = (ab)a-a = 0$. Hence $ba-e \in \operatorname{Ker} L_a$ and so $ba = e$. Thus b is the inverse of a. Also, $\lambda e-a$ is invertible iff $\lambda I - L_a$ is invertible. Hence

$$\sigma_A(a) = \sigma(L_a).$$

Although the spectrum of an operator $T \in \mathcal{B}(X)$ depends only on how T fits into the algebraic structure of $\mathcal{B}(X)$, it is of considerable interest to see how the action of T on X affects invertibility. In particular, we may distinguish the points of $\sigma(T)$ according to the reasons why $\lambda I - T$ is not invertible.

What are the obstruction to the invertibility of an operator $S \in \mathcal{B}(X)$? By the inverse-mapping theorem, S is invertible iff $\operatorname{Ker} S = (0)$ and $\operatorname{Im} S = X$. Thus if S is not invertible then either $\operatorname{Ker} S \neq (0)$ or $\operatorname{Im} S \neq X$ (or both, of course). Of these, the former is, perhaps, the more basic obstruction.

Accordingly, let us define the *point spectrum* of $T \in \mathcal{B}(X)$ as

$$\sigma_p(T) = \{\lambda \in \mathbb{C} : \operatorname{Ker}(\lambda I - T) \neq (0)\}.$$

The elements of $\sigma_p(T)$ are the *eigenvalues* of T; for an eigenvalue $\lambda \in \sigma_p(T)$, the non-zero vectors in $\operatorname{Ker}(\lambda I - T)$ are called *eigenvectors* with eigenvalue λ. Furthermore, $\operatorname{Ker}(\lambda I - T)$ is the *eigenspace* of T at λ. Clearly $\sigma_p(T) \subset \sigma(T)$.

If X is finite-dimensional and $S \in \mathcal{B}(X)$ then the two conditions $\operatorname{Ker} S = (0)$ and $\operatorname{Im} S = X$ coincide. Hence $\sigma(T) = \sigma_p(T)$ for every operator on a finite-dimensional space.

However, if X is infinite-dimensional then we may have $\operatorname{Ker} S = (0)$ and $\operatorname{Im} S \neq X$, so the point spectrum need not be the entire spectrum. More precisely, by Theorem 11.10 , $\lambda \in \sigma(T)$ iff either $\operatorname{Im}(\lambda I - T)$ is not dense in X or $\lambda I - T$ is not bounded below: there is no $\epsilon > 0$ such that $\|(\lambda I - T)x\| \geq \epsilon \|x\|$ for every $x \in X$. In the former case λ is said to belong to the *compression spectrum* $\sigma_{\text{com}}(T)$, and in the latter case, λ is said to belong to the *approximate point spectrum of T*, denoted by $\sigma_{\mathrm{ap}}(T)$. In other words,

$$\sigma_{\mathrm{ap}}(T) = \{\lambda \in \mathbb{C}: \text{there is a sequence } (x_n) \subset S(X)$$

$$\text{such that } (\lambda I - T) x_n \to 0\}.$$

Sometimes (x_n) is called an *approximate eigenvector* with eigenvalue λ. Clearly $\sigma_{\mathrm{p}}(T) \subset \sigma_{\mathrm{ap}}(T)$ and

$$\sigma(T) = \sigma_{\mathrm{ap}}(T) \cup \sigma_{\mathrm{com}}(T).$$

Sometimes the points of the spectrum are classified further: the *residual spectrum* is $\sigma_{\mathrm{r}}(T) = \sigma_{\mathrm{com}}(T) \backslash \sigma_{\mathrm{p}}(T)$ and the *continuous spectrum* is $\sigma_{\mathrm{c}}(T) = \sigma(T) \backslash (\sigma_{\mathrm{com}}(T) \cup \sigma_{\mathrm{p}}(T))$. Thus

$$\sigma(T) = \sigma_{\mathrm{p}}(T) \cup \sigma_{\mathrm{c}}(T) \cup \sigma_{\mathrm{r}}(T),$$

with the sets on the right being pairwise disjoint.

It is immediate from the definition that the approximate point spectrum $\sigma_{\mathrm{ap}}(T)$ is a closed set; the point spectrum $\sigma_{\mathrm{p}}(T)$ need not be closed.

We shall show that $\sigma(T)$ is a non-empty closed subset of the disc $\{\lambda \in \mathbb{C}: |\lambda| \leq \|T\|\}$. The latter assertion is an immediate consequence of the following simple but important result.

Theorem 1. Let $T \in \mathcal{B}(X)$ satisfy $\|T\| < 1$. Then $1 \notin \sigma(T)$ and

$$(I - T)^{-1} = \sum_{k=0}^{\infty} T^k,$$

where the series on the right is absolutely convergent in the Banach space $\mathcal{B}(X)$.

Proof. Note that

$$\sum_{k=0}^{\infty} \|T^k\| \leq \sum_{k=0}^{\infty} \|T\|^k = (1 - \|T\|)^{-1},$$

and so $\sum_{k=0}^{\infty} T^k$ is absolutely convergent.

Hence

$$(I - T) \sum_{k=0}^{\infty} T^k = (I - T) + (T - T^2) + (T^2 - T^3) + \ldots = I$$

and, similarly,

$$\left(\sum_{k=0}^{\infty} T^k\right)(I - T) = I. \qquad \square$$

Theorem 2. Suppose that $S, T \in \mathcal{B}(X)$, T is invertible and $\|S - T\| < \|T^{-1}\|^{-1}$. Then S is invertible and

$$\|S^{-1} - T^{-1}\| \leqslant \frac{\|T^{-1}\|^2 \|S - T\|}{1 - \|T^{-1}\| \|S - T\|}.$$

Proof. By Theorem 1, we have

$$[I - T^{-1}(T - S)]^{-1} = \sum_{n=0}^{\infty} [T^{-1}(T - S)]^n$$

and so

$$S^{-1} = [T - (T - S)]^{-1} = \{T[I - T^{-1}(T - S)]\}^{-1} = \sum_{n=0}^{\infty} [T^{-1}(T - S)]^n T^{-1} \tag{1}$$

and

$$\|S^{-1} - T^{-1}\| \leqslant \sum_{n=1}^{\infty} \|[T^{-1}(T - S)]^n T^{-1}\| \leqslant \frac{\|T^{-1}\|^2 \|T - S\|}{1 - \|T^{-1}\| \|T - S\|}. \qquad \square$$

Corollary 3. For a Banach space X, let

$$\mathcal{G}(X) = \{T \in \mathcal{B}(X) : T \text{ is invertible}\}.$$

Then $\mathcal{G}(X)$ is an open subset of $\mathcal{B}(X)$. Furthermore, $\mathcal{G}(X)$, with the topology inherited from $\mathcal{B}(X)$ and operator multiplication, is a topological group: it is closed under multiplication, the multiplication is jointly continuous and the map $T \mapsto T^{-1}$ is a homeomorphism of $\mathcal{G}(X)$ onto itself. $\qquad \square$

For operators of the form $\lambda_0 I - T$ and $\lambda I - T$, identity (1) takes a somewhat simpler form. Namely, if

$$|\lambda - \lambda_0| < \|(\lambda_0 I - T)^{-1}\|^{-1} = \|R(\lambda_0)\|^{-1}$$

then

$$R(\lambda) = (\lambda I - T)^{-1} = \sum_{n=0}^{\infty} (\lambda_0 - \lambda)^n (\lambda_0 I - T)^{-n-1} = \sum_{n=0}^{\infty} (\lambda_0 - \lambda)^n R(\lambda_0)^{n+1} \tag{2}$$

with the series converging in operator norm.

Theorems 1 and 2 have the following immediate consequence.

Corollary 4. The spectrum $\sigma(T)$ is a closed subset of the disc $\{\lambda \in \mathbb{C} : |\lambda| \leqslant \|T\|\}$. If $|\lambda| > \|T\|$ then

$$R(\lambda) = (\lambda I - T)^{-1} = \lambda^{-1}\left(I - \frac{T}{\lambda}\right)^{-1} = \sum_{n=0}^{\infty} \frac{T^n}{\lambda^{n+1}}. \tag{3}$$

and

$$\|R(\lambda)\| = \|(\lambda I - T)^{-1}\| \le \frac{1}{|\lambda| - \|T\|}. \qquad \square$$

Note that we do not yet know whether the spectrum $\sigma(T)$ can be empty or not. In proving that it cannot, we shall make use of Banach-space-valued analytic functions, an example of which we have just seen in (2).

Given a Banach space X and an open set $D \subset \mathbb{C}$, a function $F : D \to X$ is said to be *analytic* if for every $z_0 \in D$ there is an $r = r(z_0) > 0$ such that $D(z_0, r) = \{z \in \mathbb{C} : |z - z_0| < r\} \subset D$ and

$$F(z) = \sum_{n=0}^{\infty} a_n(z - z_0)^n \tag{4}$$

for some $a_0, a_1, \ldots \in X$ and all $z \in D(z_0, r)$, with the series in (4) being absolutely convergent. Thus (2) shows that the resolvent $R : \rho(T) \to \mathcal{B}(X)$ is an analytic function, with values in the Banach space $\mathcal{B}(X)$.

The standard results concerning analytic functions remain valid in this more general setting. For example, as we noted in Exercise 10 of Chapter 11, the analogue of Liouville's theorem holds: a norm-bounded entire function in constant. To spell it out: if $F : \mathbb{C} \to X$ is analytic and $\|F(z)\| \le M$ for some $M < \infty$ and all $z \in \mathbb{C}$ then $F(z)$ is constant. Indeed, for $f \in X^*$ the function

$$g(z) = f(F(z))$$

is a (complex-valued) analytic function and so it is constant: $g(z) = g(0)$ for all z. Hence $f(F(z) - F(0)) = 0$ for all $f \in X^*$ and so, by the Hahn–Banach theorem, $F(z) = F(0)$ for all z.

Also, the radius of convergence of (4) is $\liminf_{n\to\infty} \|a_n\|^{-1/n}$, just as in the classical case. Equivalently, the *Laurent series*

$$G(z) = \sum_{n=0}^{\infty} b_n z^{-n} \tag{5}$$

has radius of convergence $s = \limsup_{n\to\infty} \|b_n\|^{1/n}$.

Indeed, if $|z| > s$ then $(1+\epsilon)\|b_n\|^{1/n} < |z|$ for some $\epsilon > 0$ and every sufficiently large n. Hence $\|b_n z^{-n}\| < (1+\epsilon)^{-n}$ if n is sufficiently large, implying that (5) is absolutely convergent.

Conversely, if $|z| < s$ then there is an infinite sequence $n_1 < n_2 < \cdots$ such that $\|b_{n_k}\| > |z|^{n_k}$. But then $\|b_{n_k} z^{-n_k}\| > 1$ and so (5) does not converge.

The analytic functions we shall consider will take their values in $\mathcal{B}(X)$ or, more generally, in a Banach algebra. If $F_i : D_i \to A$ $(i = 1, 2)$ are analytic functions into a Banach algebra A then $F_1 F_2 : D_1 \cap D_2 \to A$ is also an analytic function. Furthermore, if

$$F_1(z) = \sum_{n=0}^{\infty} a_n (z - z_0)^n, \qquad F_2(z) = \sum_{n=0}^{\infty} b_n (z - z_0)^n \qquad \text{for } z \in D(z_0, r),$$

then

$$F_1(z) F_2(z) = \sum_{n=0}^{\infty} c_n (z - z_0)^n, \qquad \text{where } c_n = \sum_{k=0}^{n} a_k b_{n-k}.$$

These observations more than suffice to prove the main results of the chapter.

Theorem 5. The spectrum of an operator $T \in \mathcal{B}(X)$ is not empty.

Proof. Suppose that $\sigma(T) = \varnothing$. Since, as we remarked earlier, formula (2) shows that $R(\lambda)$ is an analytic function on the resolvent, which is now the entire plane, $R(\lambda)$ is an entire function. Furthermore, (3) shows that $\|R(\lambda)\| \to 0$ as $|\lambda| \to \infty$. Hence, by Liouville's theorem, $R(\lambda)$ is constant and, in fact, $R(\lambda) = 0$ for every λ. But this is clearly impossible. $\square$

For emphasis, let us put Corollary 4 and Theorems 2 and 5 together.

Theorem 6. For every complex Banach space X and operator $T \in \mathcal{B}(X)$, the spectrum $\sigma(T)$ is a non-empty closed subset of $\{\lambda \in \mathbb{C} : |\lambda| \leq \|T\|\}$. Furthermore, if $\lambda \in \rho(T)$ and

$$d(\lambda, \sigma(T)) = \min\{|\lambda - \mu| : \mu \in \sigma(T)\} = d$$

then $\|R(\lambda)\| \geq 1/d$. $\square$

It is interesting to note that Theorem 5 gives an independent proof of the fact that every $n \times n$ complex matrix has an eigenvalue. Note also that the analogue of Theorem 5 *fails* for real spaces, as shown, for example, by the rotation $(e_1, e_2) \mapsto (e_2, -e_1)$ in $\mathbb{R}^2$. The spectrum is very useful precisely because it allows techniques of complex analysis to be brought into operator theory.

From Theorem 6 it is a short step to show that the approximate point spectrum is not empty either.

Theorem 7. The approximate point spectrum $\sigma_{\mathrm{ap}}(T)$ contains the boundary $\partial\sigma(T)$ of the spectrum.

Proof. Let $\lambda \in \partial\sigma(T)$. Pick a sequence $\lambda_1, \lambda_2, \ldots \in \rho(T)$ tending to λ. Then, by the second part of Theorem 6, $\|R(\lambda_n)\| \to \infty$. Therefore there is a sequence $(y_n)_1^\infty \subset X$ such that $y_n \to 0$ and $\|R(\lambda_n)y_n\| = 1$ for every n. Setting $x_n = R(\lambda_n)y_n$, we find that $\|x_n\| = 1$ and

$$\|(\lambda I - T)x_n\| \leqslant \|(\lambda_n I - T)x_n\| + \|(\lambda - \lambda_n)x_n\| = \|y_n\| + |\lambda - \lambda_n| \to 0.$$

Hence $(x_n)_1^\infty$ is an approximate eigenvector with eigenvalue λ.   □

As an illustration of the concepts and results presented so far, let us examine the spectrum of the right shift operator S on l_p $(1 \leqslant p \leqslant \infty)$ defined by $S(x_1, x_2, \ldots) = (0, x_1, x_2, \ldots)$. Since $\|S\| = 1$, the spectrum $\sigma(S)$ is a closed non-empty subset of the closed disc $\Delta = \{\lambda \in \mathbb{C} : |\lambda| \leqslant 1\}$. Furthermore, $\|Sx\| = \|x\|$ for every x and so $\|(S-\lambda)x\| \geqslant (1-|\lambda|)\|x\|$, implying that $\sigma_{\mathrm{ap}}(S) \subset \partial\Delta = \{\lambda \in \mathbb{C} : |\lambda| = 1\}$. By Theorem 7 we have $\sigma_{\mathrm{ap}}(S) = \partial\Delta$. (Of course, this is easy to check directly.)

How much of the circle $\partial\Delta$ belongs to the point spectrum? Suppose that $Sx = \lambda x \neq 0$, where $\lambda \neq 0$. Then $0 = \lambda x_1$, $x_1 = \lambda x_2$, $x_2 = \lambda x_3$, $\ldots$, implying that $x_1 = 0$, $x_2 = 0$, $x_3 = 0$, $\ldots$. Hence $\sigma_{\mathrm{p}}(S) = \varnothing$.

Therefore the spectrum $\sigma(S)$ is the closed disc Δ, the approximate point spectrum $\sigma_{\mathrm{ap}}(S)$ is the circle $\partial\Delta$ and the point spectrum is empty.

What is the spectrum of a 'nice' function f of an operator? This is easy to answer when f is a polynomial; an analogous result holds in a much more general case, namely when f is an analytic function on an open set containing the spectrum.

Theorem 9. Let $p(t)$ be a polynomial with complex coefficients. Then for $T \in \mathscr{B}(X)$, the spectrum of $p(T)$ is precisely

$$p(\sigma(T)) = \{p(\lambda) : \lambda \in \sigma(T)\}.$$

Proof. We may assume that the leading coefficient of $p(t)$ is 1 and $p(0) = 0$. Given $\lambda_0 \in \mathbb{C}$, let

$$p(t) - \lambda_0 = \prod_{k=1}^{n} (t - \mu_k).$$

Then

$$p(T) - \lambda_0 I = \prod_{k=1}^{n} (T - \mu_k I).$$

This product fails to be invertible iff at least one of the factors, say $T - \mu_k I$, is not invertible, i.e. $\mu_k \in \sigma(T)$. Since the μ_k are the zeros of $p(t) - \lambda_0$, this happens iff $p(\mu) = \lambda_0$ for some $\mu \in \sigma(T)$.   □

The *spectral radius* of $T \in \mathcal{B}(X)$ is

$$r(T) = \sup\{|\lambda| : \lambda \in \sigma(T)\} = \max\{|\lambda| : \lambda \in \sigma(T)\}.$$

The spectral radius is a simple function of the sequence $(\|T^n\|)_1^\infty$, as shown by the following result, *Gelfand's spectral-radius formula*.

Theorem 8. For $T \in \mathcal{B}(X)$ we have

$$r(T) = \lim_{n \to \infty} \|T^n\|^{1/n}.$$

Proof. By Theorems 6 and 7 we have

$$r(T)^n = r(T^n) \leqslant \|T^n\|. \tag{6}$$

Hence $r(T) \leqslant \liminf_{n \to \infty} \|T^n\|^n$.

On the other hand, as $\rho(T) \supset \{\lambda \in \mathbb{C} : |\lambda| > r(T)\}$, relation (3) tells us that the Laurent series

$$\sum_{n=0}^{\infty} \frac{T^n}{\lambda^{n+1}}$$

is convergent for $|\lambda| > r(T)$. Hence, recalling the formula for the radius of convergence, we find that $r(T) \geqslant \limsup_{n \to \infty} \|T^n\|^{1/n}$.   □

It is easily seen that the spectral radius is an upper semicontinuous function of the operator in the norm topology; in fact,

$$r(S + T) \leqslant r(S) + \|T\|$$

for all $S, T \in \mathcal{B}(X)$ (see Exercise 8).

It is worth recalling that *all the results above are true for the spectra of elements of Banach algebras*, not only of elements of $\mathcal{B}(X)$. In the simplest of all Banach algebras, $\mathbb{C}$, every non-zero element is invertible. In fact, $\mathbb{C}$ is the *only* Banach algebra which is a division algebra.

Theorem 10 (Gelfand–Mazur theorem) Let A be a complex unital Banach algebra in which every non-zero element is invertible. Then $A = \mathbb{C}$.

Proof. Given $a \in A$, let $\lambda \in \sigma(a)$. Then $\lambda - a$ $(= \lambda I - a)$ is not invertible and so $\lambda - a = 0$, i.e. $a = \lambda$. $\square$

We know from Theorem 11.11 that $T \in \mathcal{B}(X)$ is invertible iff $T^* \in \mathcal{B}(X^*)$ is invertible. Hence $\lambda I - T$ is invertible iff $\lambda I - T^*$ is. Therefore, recalling that for a Hilbert space H, the dual H^* is identified with H by an anti-isomorphism, we get that for $T \in \mathcal{B}(H)$, $\lambda I - T$ is invertible iff $(\lambda I - T)^* = \bar{\lambda} I - T^*$ is invertible. Finally, recalling from Theorem 11.8(b) that for a normal operator T we have $\|T^n\| = \|T\|^n$ for $n \geq 1$, we have the following result.

Theorem 11.
  (a) For a Banach space X and an operator $T \in \mathcal{B}(X)$, we have $\sigma(T^*) = \sigma(T)$.
  (b) For a Hilbert space H and an operator $T \in \mathcal{B}(H)$, we have $\sigma(T^*) = \operatorname{conj} \sigma(T) = \{\bar{\lambda} : \lambda \in \sigma(T)\}$.
  (c) If T is a normal operator on a Hilbert space then $r(T) = \|T\|$. $\square$

Let us introduce another bounded subset of the complex plane associated with a linear operator. Given a Banach space X and an operator $T \in \mathcal{B}(X)$, the (*spatial*) *numerical range* of T is

$$V(T) = \{\langle Tx, f \rangle : x \in X, f \in X^*, \|x\| = \|f\| = f(x) = 1\}.$$

With the notation used before Lemma 8.10,

$$V(T) = \{f(Tx) : (x, f) \in \Pi(X)\}.$$

Thus to get a point of the numerical range, we take a point x of the unit sphere $S(X)$, a *support functional* f at x, i.e. a point of the unit sphere $S(X^*)$ taking value 1 at x, and evaluate f at Tx. It is clear that the numerical range depends on the shape of the unit ball, not only on the algebra $\mathcal{B}(X)$. If T is an operator on a Hilbert space then $V(T)$ is just the set of values taken by the hermitian form $\langle Tx, x \rangle$ on the unit sphere: $V(T) = \{\langle Tx, x \rangle : \|x\| = 1\}$. Nevertheless, the numerical range can be easier to handle than the spectrum and is often more informative. It is clear that, just as the spectrum, $V(T)$ is contained in the closed disc of centre 0 and radius $\|T\|$. Even more, the closure of $V(T)$ is sandwiched between $\sigma(T)$ and this disc. But before we show this, we prove that $V(T^*)$ can be only a little bigger than $V(T)$.

Theorem 12. For a complex Banach space X and an operator $T \in \mathcal{B}(X)$, we have

$$V(T) \subset V(T^*) \subset \overline{V(T)},$$

where $\overline{V(T)}$ is the closure of $V(T)$.

Proof. The first inclusion is easily seen since if $x \in S(X)$, $f \in S(X^*)$ and $\langle x, f \rangle = 1$ then $\hat{x} \in S(X^{**})$, $\langle f, \hat{x} \rangle = 1$ and $\langle Tx, f \rangle = \langle x, T^*f \rangle = \langle T^*f, \hat{x} \rangle$, where, as earlier, $\hat{x}$ denotes x considered as an element of X^{**}.

To see the second inclusion, let $\xi \in V(T^*)$ so that there are $f \in S(X^*)$ and $\varphi \in S(X^{**})$ such that $\langle f, \varphi \rangle = 1$ and $\xi = \langle T^*f, \varphi \rangle$. Let $0 < \epsilon < 1$. Since $B(X)$ is w^*-dense in $B(X^{**})$, there is an $x \in B(X)$ such that

$$|\langle f, \varphi - \hat{x} \rangle| < \tfrac{1}{4}\epsilon^2 \qquad \text{and} \qquad |\langle T^*f, \varphi - \hat{x} \rangle| < \tfrac{1}{4}\epsilon^2.$$

Then, by Theorem 8.11, there exist $g \in S(X^*)$ and $y \in S(X)$ such that $\langle y, g \rangle = 1$, $\|x - y\| < \epsilon$ and $\|f - g\| \le \epsilon$. Hence

$$|\langle T(x - y), f \rangle| < \epsilon\|T\| \qquad \text{and} \qquad \|\langle Ty, f - g \rangle\| \le \epsilon\|T\|.$$

But this implies that $V(T)$ has a point close to ξ, namely the point $\langle Ty, g \rangle \in V(T)$:

$$\begin{aligned}
|\langle Ty, g \rangle - \xi| &= |\langle Ty, g \rangle - \langle T^*f, \varphi \rangle| \\
&\le |\langle Ty, g \rangle - \langle T^*f, \hat{x} \rangle| + \tfrac{1}{4}\epsilon^2 \\
&\le |\langle Ty, g \rangle - \langle Ty, f \rangle| + \epsilon\|T\| + \tfrac{1}{2}\epsilon^2 \\
&\le 2\epsilon\|T\| + \tfrac{1}{4}\epsilon^2. \qquad\qquad \square
\end{aligned}$$

Theorem 13. For a complex Banach space X, the spectrum of T is contained in the closure $\overline{V(T)}$ of the numerical range $V(T)$.

Proof. Suppose that

$$d(\lambda, V(T)) = \inf\{|\lambda - \mu| : \mu \in V(T)\} = d > 0.$$

By Theorem 12 we also have $d(\lambda, V(T^*)) = d$. To prove that $\sigma(T) \subset \overline{V(T)}$, we have to show that $\lambda I - T$ is invertible.

Given $x \in S(X)$, pick a support functional $f \in S(X^*)$ at x, i.e. a norm-1 functional with $\langle x, f \rangle = 1$. Then $\langle Tx, f \rangle \in V(T)$ and so

$$\|(\lambda I - T)x\| \ge |\langle (\lambda I - T)x, f \rangle| = |\lambda - \langle Tx, f \rangle| \ge d.$$

Hence

$$\|(\lambda I - T)x\| \ge d\|x\|$$

for all $x \in X$.

Similarly, as $d(\lambda, V(T^*)) = d$, we have

$$\|(\lambda I - T^*)f\| \ge d\|f\|$$

for all $f \in X^*$. Thus both $\lambda I - T$ and $(\lambda I - T)^*$ are bounded below. But then, by Theorem 11.12, $\lambda I - T$ is invertible.   □

Another aspect of the connection between the spectrum and the numerical range given in Theorem 13 is that $\operatorname{co}\sigma(T)$, the convex hull of spectrum, is precisely $\bigcap \overline{\operatorname{co}}\, V(T)$, where the intersection is taken over all numerical ranges $V(T)$ with respect to norms on X equivalent to the given norm. But we shall not give a proof of this result.

The rest of the chapter is about a striking application of the spectral-radius formula to obtain a remarkable theorem related to material beyond the main body of this book. The theorem is Johnson's uniqueness-of-norm theorem, but the beautiful and unexpectedly simple proof we present is due to Ransford.

Let us start with a classical inequality concerning complex functions, namely Hadamard's three-circles theorem, stating that if f is analytic in the annulus $R_1 < |z| < R_2$ then $M_r = \max_{|z|=r}|f(z)|$ is a convex function of $\log r$ for $R_1 < r < R_2$. Thus if $f(z)$ is analytic in the open disc $|z| < R_0$, where $R_0 > 1$, then

$$|f(1)|^2 \le \max_{|z|=R}|f(z)|\, \max_{|z|=1/R}|f(z)| \tag{7}$$

for all R $(1 < R < R_0)$.

To see (7), note simply that $g(z) = f(z)f(1/z)$ is analytic in (an open set containing) the annulus $1/R \le |z| \le R$ and so attains its maximum there on the boundary. In particular,

$$|f(1)|^2 = |g(1)| \le \max_{|z|=R}|g(z)| \le \max_{|z|=1/R}|f(z)|.$$

A similar inequality holds for Banach-space-valued analytic functions, with the norm replacing the modulus. In particular, if $f(z)$ is analytic in the open disc $|z| < R_0$, with values in a Banach space X then

$$\|f(1)\|^2 \le \max_{|z|=R}\|f(z)\|\, \max_{|z|=1/R}\|f(z)\|. \tag{8}$$

for all R $(1 < R < R_0)$. Indeed, let $f(z) = \sum_{n=0}^{\infty} a_n z^n$ $(a_n \in X)$ and let $\varphi \in S(X^*)$ be a support functional at $f(1) = \sum_{n=0}^{\infty} a_n$:

$$\varphi\left(\sum_{n=0}^{\infty} a_n\right) = \sum_{n=0}^{\infty} \varphi(a_n) = \left\|\sum_{n=0}^{\infty} a_n\right\|.$$

Set

$$g(z) = \varphi(f(z)) = \sum_{n=0}^{\infty} \varphi(a_n) z^n.$$

Then, by (7)

$$\|f(1)\|^2 = |g(1)|^2 \leq \max_{|z|=R} |g(z)| \max_{|z|=1/R} |g(z)|$$

$$= \max_{|z|=R} |\varphi(f(z))| \max_{|z|=1/R} |\varphi(f(z))|$$

$$= \max_{|z|=R} \|f(z)\| \max_{|z|=1/R} \|f(z)\|,$$

proving (8).

In fact, if we have an analytic function with values in a Banach *algebra* then in inequality (8) we may replace the norm by the spectral radius.

Lemma 13. Let $f(z)$ be an analytic function in the open disc $|z| < R_0$, with values in a Banach algebra A. Then

$$r(f(1))^2 \leq \max_{|z|=R} r(f(z)) \max_{|z|=1/R} r(f(z)).$$

Proof. From the spectral-radius formula, we know that $\|f(z)^{2^n}\|^{2^{-n}}$ is monotone decreasing to $r(f(z))$, and $r(f(z))$ is a continuous function of z for $|z| = R$ and $|z| = 1/R$. Consequently, by Dini's theorem (Theorem 6.5), for every $\epsilon > 0$ there is an n such that

$$\max_{|z|=R} \|f(z)^{2^n}\|^{2^{-n}} \max_{|z|=1/R} \|f(z)^{2^n}\|^{2^{-n}} \leq \max_{|z|=R} \{r(f(z))+\epsilon\} \max_{|z|=1/R} \{r(f(z))+\epsilon\}.$$

Applying Theorem 9 and inequality (8), we find that

$$r(f(1))^2 \leq \|f(z)^{2^n}\|^{2^{-n}}$$

$$\leq \max_{|z|=R} \|f(z)^{2^n}\|^{2^{-n}} \max_{|z|=1/R} \|f(z)^{2^n}\|^{2^{-n}}$$

$$\leq \max_{|z|=R} \{r(f(z))+\epsilon\} \max_{|z|=1/R} \{r(f(z))+\epsilon\}.$$

As this holds for every $\epsilon > 0$, the result follows. $\qquad\square$

The *radical* Rad B of a complex (unital) Banach algebra B is the intersection of all the maximal left ideals of B. The following lemma relates the radical to the spectral radius.

Lemma 15. If $b \in B$ is such that $r(b'b) = 0$ for all $b' \in B$ then $b \in \text{Rad } B$.

Proof Suppose that $b \notin \operatorname{Rad} B$, that is $b \notin L$ for some maximal left ideal L. Then $Bb + L$ is a left ideal properly containing L, and so $Bb + L = B$. Hence $e = b'b + l$ for some $b' \in B$ and $l \in L$, where e is the identity. But then $e - b'b = l \in L$ and so $e - b'b$ is not invertible. Therefore $r(b'b) \geq 1$. □

Now we are ready to give Ransford's proof of Johnson's theorem, which is slightly more than the assertion that if $\operatorname{Rad} B = \{0\}$ then all Banach-algebra norms on B are equivalent.

Theorem 16. Let A and B be Banach algebras, with $\operatorname{Rad} B = \{0\}$. Then every surjective homomorphism $\theta : A \to B$ is automatically continuous.

Proof. Suppose that $a_n \to 0$ in A and $\theta(a_n) \to b$ in B. By the closed-graph theorem (Theorem 5.8) it suffices to show that $b = 0$. Since $\operatorname{Rad} B = \{0\}$, this is the same as showing that $b \in \operatorname{Rad} B$, and by Lemma 15 this follows if we show that $r(b'b) = 0$ for all $b' \in B$.

Let then $b' \in B$. Pick $a, a' \in A$ with $\theta(a) = b$ and $\theta(a') = b'$. Set $c_n = a'a_n$, $c = a'a$, $d_n = \theta(c_n)$ and $d = \theta(c) = b'b$. Then $c_n \to 0$ in A and $d_n = \theta(c_n) \to d$ in B. Define a linear function $p_n : \mathbb{C} \to A$ by

$$p_n(z) = c_n z + (c - c_n)$$

and set

$$q_n(z) = \theta(p_n(z)) = \theta(c_n) z + [\theta(c) - \theta(c_n)].$$

Note that $p_n(1) = c$ and $q_n(1) = d$.

Since a homomorphism does not increase the spectral radius.

$$r(q_n(z)) \leq \min\{\|p_n(z)\|, \|q_n(z)\|\}$$
$$\leq \min\{|z|\|c_n\| + \|c - c_n\|, |z|\|d_n\| + \|d - d_n\|\}.$$

Hence, by Lemma 14, for all $n \geq 1$ and $R > 1$ we have

$$r(d)^2 \leq (R\|c_n\| + \|c - c_n\|)(R^{-1}\|d_n\| + \|d - d_n\|).$$

Letting $n \to \infty$, we find that

$$r(d)^2 \leq \|c\|(R^{-1}\|d\|)$$

and so, letting $R \to \infty$, we see that $r(d) = r(b'b) = 0$, as desired. □

Corollary 17. (Johnson's uniqueness-of-norm theorem) Let B be a complex unital Banach algebra with norm $\|\cdot\|_0$ and $\operatorname{Rad} B = \{0\}$. Then every Banach algebra norm on B is equivalent to $\|\cdot\|_0$. □

The algebra $\mathcal{B}(X)$ of bounded linear operators on a Banach space X satisfies the conditions in Corollary 17, so all norms on $\mathcal{B}(X)$ turning it into a Banach algebra are equivalent.

Exercises

In the exercises below, X is a complex Banach space and $T \in \mathcal{B}(X)$.

1. We call T a *left (right) divisor of zero* if there is an $S \in \mathcal{B}(X)$ such that $S \neq 0$ and $TS = 0$ $(ST = 0)$. Show that the point spectrum of T is

$$\sigma_p(T) = \{\lambda \in \mathbb{C} : \lambda I - T \text{ is a left divisor of zero}\}$$

and the compression spectrum is

$$\sigma_{com}(T) = \{\lambda \in \mathbb{C} : \lambda I - T \text{ is a right divisor of zero}\}.$$

2. We call T a *left (right) topological divisor of zero* if there are $T_1, T_2, \ldots \in \mathcal{B}(X)$ such that $\|T_n\| = 1$ and $TT_n \to 0$ $(T_n T \to 0)$. Show that the approximate point spectrum of T is

$$\sigma_{ap}(T) = \{\lambda \in \mathbb{C} : \lambda I - T \text{ is a left topological divisor of zero}\}.$$

3. Show that T is a right topological divisor of zero iff $T^* \in \mathcal{B}(X^*)$ is a left topological divisor of zero.

4. Show that if $\lambda \in \sigma(T)$ then $\lambda I - T$ is either a right divisor of zero or a left topological divisor of zero. Deduce that

$$\sigma(T) = \sigma_{com}(T) \cup \sigma_{ap}(T).$$

5. Let X be reflexive. Prove that if T is not invertible and is neither a left nor a right divisor of zero then it is both a left topological divisor of zero and a right topological divisor of zero.

6. Show that

$$\sigma_{ap}(T) = \{\lambda \in \mathbb{C} : \lambda I - T^* \text{ is not surjective}\}$$

and

$$\sigma_{ap}(T^*) = \{\lambda \in \mathbb{C} : \lambda I - T \text{ is not surjective}\}.$$

7. Suppose that $S, T \in \mathcal{B}(X)$ commute: $ST = TS$. Show that

$$r(S + T) \leq r(S) + r(T) \qquad \text{and} \qquad r(ST) \leq r(S) r(T).$$

Show also that these inequalities need not hold if S and T are not assumed to commute.

8. Show that for $S, T \in \mathcal{B}(X)$ we have

$$r(S + T) \leqslant r(S) + \|T\|.$$

Show also that if $r(S) = 0$ then $r : \mathcal{B}(X) \to [0, \infty)$ is continuous at S.

9. Suppose that $r(T) < 1$. Show that

$$\|x\|_0 = \sum_{n=0}^{\infty} \|T^n x\|$$

is a norm on X, equivalent to the original norm $\|\cdot\|$.

10. Prove that

$$r(T) = \inf\{\|T\|' : \|\cdot\|' \text{ is a norm on } X, \text{ equivalent to } \|\cdot\|\}.$$

11. Check that the resolvent $R(\lambda)$ of T satisfies the *resolvent identity*

$$R(\lambda) - R(\mu) = (\mu - \lambda) R(\lambda) R(\mu) = (\mu - \lambda) R(\mu) R(\lambda)$$

for all $\lambda, \mu \in \rho(T)$.

12. Check that if $S, T \in \mathcal{B}(X)$, $\lambda \in \rho(ST)$ and $\lambda \neq 0$, then $\lambda \in \rho(TS)$ and

$$(\lambda I - Ts)^{-1} = \lambda^{-1} + \lambda^{-1} T (\lambda - ST)^{-1} S.$$

Deduce that $\sigma(ST) \cup \{0\} = \sigma(TS) \cup \{0\}$. Show also that $\sigma(ST) = \sigma(TS)$ need not hold.

13. Let K be a non-empty compact subset of $\mathbb{C}$. Show that K is the spectrum of a Hilbert space operator: there is an operator $S \in \mathcal{B}(H)$, where H is a Hilbert space, such that $\sigma(S) = K$.

14. Show that if T is a normal operator on a Hilbert space then $r(T) = \|T\|$.

15. For $w = (w_k)_1^{\infty} \in l_\infty$ define $T_w : l_2 \to l_2$ by $T_w(x) = (w_k x_k)_{k=1}^{\infty}$. Show that $\|T_w\| = \|w\|_\infty = \sup_k |w_k|$. Show also that eigenvalues of T are $w_1, w_2, \ldots$ and $\sigma(T) = \overline{\{w_n\}}$. What is T_w^*?

16. For $w = (w_k)_1^{\infty} \in l_\infty$ define $S_w : l_2 \to l_2$ by

$$S_w((x_1, x_2, \ldots)) = (0, w_1 x_1, w_2 x_2, \ldots).$$

Express $\|S_w\|$ and the spectral radius $r(S_w)$ in terms of the sequence w.

17. Let $\Delta = \{z \in \mathbb{C} : |z| \leqslant 1\}$ and $H = L_2(\Delta)$. Let $T \in \mathcal{B}(H)$ be the operator of multiplication by $\bar{z}$. Show that $\sigma(T) = \Delta$ and T has no eigenvalue.

18. Use the Hahn–Banach theorem to show that for $T \in \mathcal{B}(X)$ we have

$$\sup \operatorname{Re} V(T) = \sup\{c \in \mathbb{R}: \text{there is an } x = x(c) \in X \ (x \neq 0),$$

$$\text{such that } \|(1 - rc + rT)x\| \geq \|x\| \text{for all } r \geq 0\}.$$

19. An operator $T \in \mathcal{B}(X)$ is said to be *dissipative* if $\sup \operatorname{Re} V(T) \leq 0$. Show that if T is dissipative then

$$\|x - rTx\| \geq \|x\|$$

for all $x \in X$ and $r \geq 0$.

20. Show that if $\lambda \in \partial\overline{\operatorname{co}}\, V(T)$ and $(\lambda I - T)x = 0$ then

$$\|x + (\lambda I - T)y\| \geq \|x\|$$

for all $y \in X$. (Note that if T is dissipative and $Tx = 0$ then

$$\left\|x - \frac{y}{r} + \frac{1}{r}T\left(x - \frac{y}{r}\right)\right\| \geq \left\|x - \frac{y}{r}\right\|$$

for $r > 0$.) Deduce that if $\lambda \in \partial\overline{\operatorname{co}}\, V(T)$ and $(\lambda I - T)^2 z = 0$ then $(\lambda I - T)z = 0$.

21. An operator $T \in \mathcal{B}(X)$ is said to be *hermitian* if $V(T) \subset \mathbb{R}$. (Note that X is a *Banach space*; for a Hilbert space this definition coincides with the usual (earlier) definition of a hermitian operator on a Hilbert space.) Thus T is hermitian iff both iT and $-iT$ are dissipative. Prove that T is hermitian iff

$$\|\exp(iT)x\| = \|x\|$$

for all $r \in \mathbb{R}$ and $x \in X$, where

$$\exp S = \sum_{k=0}^{\infty} \frac{S^k}{k!}$$

for $S \in \mathcal{B}(X)$.

22. Let H be a Hilbert space and let $S \subset \mathcal{B}(H)$ be such that $V(S) \subset [0, \infty)$. Show that

$$\langle Sx, y \rangle \leq \langle Sx, x \rangle^{1/2} \langle Sy, y \rangle^{1/2}$$

and deduce that

$$\|Sx\|^2 \leq \langle Sx, x \rangle \|S\|$$

for all $x, y \in H$. Deduce that if $0 \in \overline{V(S)}$ then $0 \in \sigma(S)$.

23. Show that if S is a hermitian operator on a Hilbert space then

$$\overline{V(S)} = \operatorname{co}\sigma(S).$$

(Note that the assertions in the last two exercises are easily deduced from Theorem 11 (c) as well.)

24. Prove that the result in the previous exercise holds for a normal operator S on a Hilbert space.

25. Let M be a proper ideal of a unital Banach algebra B. Show that the closure of M is also a proper ideal. Deduce that every maximal ideal of B is closed.

26. Let M be a maximal ideal of a complex unital Banach algebra B. Show that B/M is also a complex unital Banach algebra.

The aim of the next five exercises is to prove the commutative Gelfand–Naĭmark theorem. In these exercises A is a commutative complex unital Banach algebra.

27. Show that if M is a maximal ideal of A then A/M is isometrically isomorphic to $\mathbb{C}$.

28. Let $h : A \to \mathbb{C}$ be a non-zero homomorphism. Show that $\|h\| = 1$.

29. Let $\mathcal{M}$ be the set of maximal ideals of A. By Exercise 27, $\mathcal{M}$ may be identified with the set of non-zero homomorphisms $h : A \to \mathbb{C}$. By Exercise 28, this is a subset of $B(A^*)$. Give $\mathcal{M}$ the relative weak* topology (i.e. the topology induced by the weak* topology on A^*). Endowed with this topology, we call $\mathcal{M}$ the *maximal ideal space* of A. Prove that $\mathcal{M}$ is a compact Hausdorff space.

30. For $x \in A$ the *Gelfand transform* of x is the function $\hat{x} : \mathcal{M} \to \mathbb{C}$ defined by $\hat{x}(h) = h(x)$, where $h \in \mathcal{M}$ is considered as a homomorphism $h : A \to \mathbb{C}$. Show that the spectrum $\sigma(x)$ is the range of $\hat{x}$.

31. Show that if A is a commutative unital C*-algebra with maximal ideal space $\mathcal{M}$ then the Gelfand transformation $x \mapsto \hat{x}$ maps A isometrically and isomorphically onto $C(\mathcal{M})$, the commutative C*-algebra of continuous functions on the compact Hausdorff space $\mathcal{M}$.

32. Let $l_1(\mathbb{Z})$ be the algebra of all (doubly infinite) complex sequences $x = (x_n)_{-\infty}^{\infty}$ such that

$$\|x\|_1 = \sum_{n=-\infty}^{\infty} |x_n| < \infty,$$

with convolution product $x*y = z$, where

$$z_n = \sum_{k=-\infty}^{\infty} x_k y_{n-k}.$$

Show that $l_1(\mathbb{Z})$ is a commutative Banach algebra. Show also that the maximal ideal space of $l_1(\mathbb{Z})$ can be identified with the unit circle $\mathbb{T} = \{z \in \mathbb{C}: |z| = 1\}$, where $z: l_1(\mathbb{Z}) \to \mathbb{C}$ is defined by

$$z(x) = \sum_{n=-\infty}^{\infty} x_n z^n.$$

33. Deduce from the result in the previous exercises that if

$$f(t) = \sum_{n=-\infty}^{\infty} x_n e^{int} \neq 0$$

for all t, and $\sum_{n=-\infty}^{\infty} |x_n| < \infty$, then

$$\frac{1}{f(t)} = \sum_{n=-\infty}^{\infty} y_n e^{int}, \qquad \text{where } \sum_{n=-\infty}^{\infty} |y_n| < \infty.$$

34$^+$. Let H be a Hilbert space and let $T \in \mathcal{B}(H)$. Show that for every $\epsilon > 0$ there is an invertible operator $S \in \mathcal{B}(H)$ such that $\|STS^{-1}\| < r(T) + \epsilon$.

Notes

The original reference to Gelfand's spectral-radius formula is I. M. Gelfand, *Normierte Ringe*, Mat. Sbornik N. S., **9** (51) (1941), 3–24; this is also one of the references to the Gelfand–Mazur theorem; the other is S. Mazur, *Sur les anneaux linéaires*, C. R. Acad. Sci. Paris, **207** (1938), 1025–27.

Theorem 12 is from B. Bollobás, *An extension to a theorem of Bishop and Phelps*, Bull. London Math. Soc., **2** (1970), 181–2, and Theorem 13 is from J. P. Williams, *Spectra of products and numerical ranges*, J. Math. Anal. and Appl., **17** (1967), 214–20. A good account of numerical ranges can be found in F. F. Bonsall and J. Duncan, *Numerical Ranges II*, London Mathematical Society Lecture Note Series, vol. 10, Cambridge University Press, 1973, vii + 179 pp.

Johnson's theorem is from B. E. Johnson, *The uniqueness of the (complete) norm topology*, Bull. Amer. Math. Soc., **73** (1967), 537–9; its simple proof is from T. J. Ransford, *A short proof of Johnson's uniqueness-of-norm theorem*, Bull. London Math. Soc., **21** (1989), 487–8.

The commutative Gelfand–Naĭmark theorem is taken from I. M. Gelfand and M. A. Naĭmark, *On the embedding of normed rings into the ring of operators in Hilbert space*, Mat. Sbornik, **12** (1943), 197–213; another classical reference is M. A. Naĭmark, *Normed Rings*, Revised English edition; translated from the Russian by Leo F. Boron, Groningen: Noordhoff, 1964.

13. COMPACT OPERATORS ON BANACH SPACES

For the sake of simplicity we shall assume that all the spaces appearing in this chapter are *complex Banach spaces*. Our aim is to study a class of operators closely resembling the operators on finite-dimensional spaces; we shall show that these operators are somewhat similar to the $n \times n$ complex matrices.

Vaguely speaking, the operators we shall look at are 'small' in the sense that they map the unit ball into a 'small' set. To be precise, an operator $T \in \mathcal{B}(X, Y)$ is *compact* if TB_X, the image of the unit ball B_X under T, is a relatively compact (i.e. totally bounded) subset of Y. Thus T is compact if and only if $\overline{TB_X}$ is compact. Equivalently, T is compact if and only if for every bounded sequence $(x_n) \subset X$ the sequence (Tx_n) has a convergent subsequence.

We shall write $\mathcal{B}_0(X, Y)$ for the set of compact operators from X into Y. Analogously to $\mathcal{B}(X)$, we write $\mathcal{B}_0(X)$ for $\mathcal{B}_0(X, X)$.

Examples 1. (i) Every *finite rank operator* $T \in \mathcal{B}(X, Y)$ is compact, i.e. if $\dim \operatorname{Im} T = \dim TX < \infty$ then $T \in \mathcal{B}_0(X, Y)$. Indeed, set $Z = \operatorname{Im} T$. Since Z is finite-dimensional, B_Z is compact and so TB_X is a subset of the compact set $\|T\|B_Z$.

We shall denote by $\mathcal{B}_{00}(X, Y)$ the set of (bounded) *finite rank operators* from X to Y.

(ii) Every bounded linear functional $f \in X^*$ is a compact operator from X to $\mathbb{C}$.

(iii) Let I be the closed unit interval $[0, 1]$ and let X be the Banach space $C(I)$ of continuous functions with the supremum norm. Let $K(x, y) \in C(I \times I)$, i.e. let K be a continuous function on the closed unit square $I \times I$. For $f \in C(I)$ define a function $Tf \in C(I)$ by

$$(Tf)(x) = \int_0^1 K(x,y)f(y) \, dy.$$

Then $T \in \mathscr{B}(X)$ and it is easily seen that, in fact, $T \in \mathscr{B}_0(X)$. Indeed, by the Arzelá–Ascoli theorem (Theorem 6.4) we have only to check that TB_X is uniformly bounded and equicontinuous. If $|K(x,y)| \le N$ for every $(x,y) \in I \times I$ then

$$|(Tf)(x)| \le N \int_0^1 |f(y)| \, dy \le N\|f\|,$$

and so TB_X is uniformly bounded by N. Also, for $\epsilon > 0$, there is a $\delta > 0$ such that if $|x_1 - x_2| < \delta$ then $|K(x_1,y) - K(x_2,y)| < \epsilon$. Therefore if $f \in B_X$ and $|x_1 - x_2| < \delta$ then

$$|(Tf)(x_1) - (Tf)(x_2)| \le \int_0^1 |K(x_1,y) - K(x_2,y)| \, |f(y)| \, dy$$

$$\le \int_0^1 |K(x_1,y) - K(x_2,y)| \, dy < \epsilon.$$

Thus TB_X is indeed equicontinuous and so T is compact, as claimed.

(iv) Let H be a Hilbert space with orthonormal bases $(e_i)_1^\infty$ and $(f_j)_1^\infty$. For every operator $T \in \mathscr{B}(H)$ we have

$$\sum_{i=1}^\infty \|Te_i\|^2 = \sum_{i=1}^\infty \sum_{j=1}^\infty |\langle Te_i, e_j \rangle|^2 = \sum_{i=1}^\infty \sum_{j=1}^\infty |\langle e_i, T^*e_j \rangle|^2 = \sum_{j=1}^\infty \|T^*e_j\|^2$$

and so

$$\sum_{i=1}^\infty \|Tf_i\|^2 = \sum_{i=1}^\infty \sum_{j=1}^\infty |\langle Tf_i, e_j \rangle|^2 = \sum_{j=1}^\infty \|T^*e_j\|^2 = \sum_{i=1}^\infty \|Te_i\|^2.$$

This shows that

$$\|T\|_{\mathrm{HS}} = \left(\sum_{i=1}^\infty \|Te_i\|^2 \right)^{1/2}$$

is independent of the orthonormal basis $(e_i)_1^\infty$.

An operator $T \in \mathscr{B}(H)$ is said to be a *Hilbert–Schmidt operator* if its *Hilbert–Schmidt norm*, $\|T\|_{\mathrm{HS}}$, is finite. One often writes $\mathscr{B}_2(H)$ for the set of Hilbert–Schmidt operators on H, and $\|T\|_2$ for $\|T\|_{\mathrm{HS}}$.

As the notation $\|T\|_2$ indicates, the Hilbert–Schmidt norm is the l_2-norm of the sequence formed by the entries of the matrix representation of T. Indeed, set $a_{ij} = \langle Te_j, e_i \rangle$ so that T is given by the matrix $A = (a_{ij})$ in the sense that

$$T\left(\sum_{i=1}^{\infty} x_i e_i\right) = \sum_{i=1}^{\infty} \left(\sum_{j=1}^{\infty} a_{ij} x_j\right) e_i.$$

Then

$$\|T\|_{\mathrm{HS}} = \|T\|_2 = \left(\sum_{i=1}^{\infty} \sum_{j=1}^{\infty} |a_{ij}|^2\right)^{1/2}.$$

Putting it another way, with

$$a_i = \sum_{j=1}^{\infty} \bar{a}_{ij} e_j$$

we have $a_i \in H$ for every i,

$$\|T\|_{\mathrm{HS}} = \left(\sum_{i=1}^{\infty} \|a_i\|^2\right)^{1/2} \quad \text{and} \quad Tx = \sum_{i=1}^{\infty} \langle x, a_i \rangle e_i.$$

In particular,

$$\|Tx\|^2 = \sum_{i=1}^{\infty} |\langle x, a_i \rangle|^2 \leq \sum_{i=1}^{\infty} \|x\|^2 \|a_i\|^2 = \|x\|^2 \|T\|_{\mathrm{HS}}^2,$$

and so $\|T\| \leq \|T\|_{\mathrm{HS}}$.

It is easily seen that every Hilbert–Schmidt operator is compact. Indeed, with the notation as above, put

$$A = \left\{x = \sum_{i=1}^{\infty} x_i e_i \in H : |x_i| \leq \|a_i\|, i = 1, 2, \ldots\right\}.$$

Since $\sum_{i=1}^{\infty} \|a_i\|^2 < \infty$, the set A is a compact subset of H (see Exercise 1). Since TB_H is a subset of A, it is relatively compact.   □

The class of compact operators is an example of a closed operator ideal. An *operator ideal* $\mathcal{R}$ is a function that assigns to every pair X, Y of Banach spaces a subset $\mathcal{R}(X, Y)$ of $\mathcal{B}(X, Y)$ such that if $T \in \mathcal{R}(X, Y)$, $S \in \mathcal{B}(Y, Z)$ and $R \in \mathcal{B}(W, X)$ then $ST \in \mathcal{R}(X, Z)$ and $TR \in \mathcal{R}(W, Y)$.

Theorem 2.

(a) $\mathcal{B}_0(X, Y)$ is a closed subspace of $\mathcal{B}(X, Y)$.

(b) If $T \in \mathcal{B}_0(X, Y)$, $S \in \mathcal{B}(Y, Z)$ and $R \in \mathcal{B}(W, X)$ then $ST \in \mathcal{B}_0(X, Z)$ and $TR \in \mathcal{B}_0(W, Y)$.

Proof. (a) Let us show first that $\mathcal{B}_0(X, Y)$ is a subspace of $\mathcal{B}(X, Y)$, i.e. if $S, T \in \mathcal{B}_0(X, Y)$ and $\lambda, \mu \in \mathbb{C}$ then $\lambda S + \mu T \in \mathcal{B}_0(X, Y)$. Let (x_n) be a bounded sequence in X. As S is compact, (x_n) has a subsequence, say (x_{n_k}), such that (Sx_{n_k}) is convergent. The operator T is also compact, and so (x_{n_k}) has a subsequence, say (x_{m_k}), such that (Tx_{m_k}) is convergent. But then $((\lambda S + \mu T)x_{m_k})$ is also a convergent sequence.

Now let us show that $\mathcal{B}_0(X, Y)$ is a closed subset of $\mathcal{B}(X, Y)$. Suppose $T_n \in \mathcal{B}_0(X, Y)$ and $T_n \to T \in \mathcal{B}(X, Y)$. We have to show that TB_X is totally bounded. Given $\epsilon > 0$ let n be such that $\|T_n - T\| < \epsilon$. As T_n is compact, there are $x_1, \ldots, x_m \in B_X$ such that $\{T_n x_i : 1 \leqslant i \leqslant m\}$ is an ϵ-net in $T_n B_X$. Thus if $x \in B_X$ then there exists an x_i such that $\|T_n x - T_n x_i\| < \epsilon$. Then

$$\|Tx - Tx_i\| \leqslant \|(T - T_n)x\| + \|T_n x - T_n x_i\| + \|(T_n - T)x_i\| < 3\epsilon,$$

showing that $\{Tx_i : 1 \leqslant i \leqslant m\}$ is a 3ϵ-net in TB_X.

(b) Note that a bounded linear operator maps a bounded sequence into a bounded sequence and a convergent sequence into a convergent sequence. $\qquad\square$

Since finite rank operators are compact, by Theorem 2 (a) every limit of finite rank operators is compact. The problem of whether every compact operator can be obtained in this way was, for many years, one of the best-known problems in functional analysis. After about 40 years the *approximation problem* was solved in the negative by Per Enflo in 1973, who constructed a separable reflexive Banach space X for which $\mathcal{B}_0(X)$ is not the closure of $\mathcal{B}_{00}(X)$. Since $\mathcal{B}_0(X)$ *is* the closure of $\mathcal{B}_{00}(X)$ whenever X has a (Schauder) basis (see Exercise 7), Enflo's example also showed that not every separable reflexive Banach space has a basis, solving another long-standing question. As so often in mathematics, the counterexample turned out to be the *start* of the story: it opened up whole new fields of research on approximation and basis problems. But we cannot go into that in this book.

The operator ideal of compact operators is closed under taking adjoints as well.

Theorem 3. An operator $T \in \mathcal{B}(X, Y)$ is compact if and only if $T^* \in \mathcal{B}(Y^*, X^*)$ is compact.

Proof. (i) Suppose first that $T \in \mathcal{B}(X, Y)$ is compact, i.e. the set $K = \overline{T}B_X$ is compact. For a functional $f \in Y^*$ let Rf be the restriction of f to K. Clearly $Rf \in C(K)$ and, in fact, the map $R : Y^* \to C(K)$ is a bounded linear map, where, as usual, $C(K)$ is taken with the supremum norm. Let $\Phi = RB_{Y^*}$. Note that for $f \in Y^*$ we have

$$\|T^*f\| = \sup\{\langle x, T^*f\rangle : x \in B_X\}$$
$$= \sup\{\langle Tx, f\rangle : x \in B_X\}$$
$$= \sup\{\langle y, f\rangle : y \in TB_X\}$$
$$= \sup\{\langle y, f\rangle : y \in K\} = \|Rf\|.$$

This shows that $T^*B_{Y^*}$ is isometric to $\Phi \subset C(K)$, with the isometry given by $T^*f \mapsto Rf$.

Consequently $T^*B_{Y^*}$ is totally bounded if and only if Φ is. By the Arzelá–Ascoli theorem, Φ is totally bounded if and only if it is uniformly bounded and equicontinuous. Both conditions are easily checked.

If $f \in B_{Y^*}$ then $\|T^*f\| \leq \|T^*\| = \|T\|$, and so Φ is uniformly bounded. Also, if $f \in B_{Y^*}$ and $y, y' \in K$ then

$$|(Rf)(y) - (Rf)(y')| = |f(y - y')| \leq \|y - y'\|,$$

and so Φ is equicontinuous.

(ii) Now suppose that $T^* \in \mathcal{B}(Y^*, X^*)$ is compact. By part (i), the map $T^{**} \in \mathcal{B}(X^{**}, Y^{**})$ is compact, i.e. $T^{**}B_{X^{**}}$ is relatively compact. But under the natural embeddings $X \subset X^{**}$ and $Y \subset Y^{**}$ we have $TB_X = T^{**}B_X \subset T^{**}B_{X^{**}}$, and so TB_X is also relatively compact. $\quad\square$

Theorems 2 and 3 state that the compact operators form a closed operator ideal that is also closed under taking adjoints. In particular, for a Banach space X, the set $\mathcal{B}_0(X)$ is a closed ideal of the Banach algebra $\mathcal{B}(X)$. If H is a Hilbert space then $\mathcal{B}_0(H)$ is a closed ideal of H which is also closed under taking adjoints: it is a closed *-ideal.

The main aim of the chapter is to present the spectral theory of compact operators, due to Frigyes (Friedrich or Frédéric) Riesz. Recall that the spectrum of an operator $T \in \mathcal{B}(X)$ is

$$\sigma(T) = \{\lambda \in \mathbb{C} : T - \lambda I \text{ does not have a bounded inverse}\},$$

where I is the identity operator on X. We proved in the previous chapter that for every $T \in \mathcal{B}(X)$ the spectrum of T is a non-empty closed subset of $\mathbb{C}$, contained in the disc $\{z \in \mathbb{C} : |z| \leq \|T\|\}$. As we shall see, for a compact operator $T \in \mathcal{B}(X)$, the spectrum of T resembles the spectrum of an operator on a finite-dimensional space, i.e. the spectrum of an $n \times n$ matrix. To be precise, if X is infinite-dimensional and $T \in \mathcal{B}_0(X)$ then $\sigma(T)$ is a countable set whose only accumulation point is 0 and if $\lambda \in \sigma(T)$ $(\lambda \neq 0)$ then λ is an eigenvalue of T with finitely many linearly independent eigenvectors.

Theorem 4. Let $T \in \mathcal{B}_0(X)$ and $a > 0$. Then T has only finitely many linearly independent eigenvectors with eigenvalues having modulus at least a.

Proof. Suppose $x_1, x_2, \ldots$ is an infinite sequence of linearly independent eigenvectors such that $Tx_i = \lambda_i x_i \neq 0$ and $|\lambda_i| \geq a$ for every i. Set $X_n = \lin\{x_1, \ldots, x_n\}$. By Theorem 4.8(b), there exists a sequence $(y_n)_1^\infty \subset X$ such that $y_n \in X_n$ and $d(y_n, X_{n-1}) = \|y_n\| = 1$.

Put $z_n = y_n/\lambda_n$ and note that $\|z_n\| \leq 1/a$, $Tz_n \in X_n$ and $Tz_n - y_n \in X_{n-1}$. Indeed, the first two assertions are obvious and if $y_n = \sum_{k=1}^n c_k x_k$ then

$$Tz_n - y_n = \sum_{k=1}^{n-1} \left(\frac{\lambda_k}{\lambda_n} - 1\right) c_k x_k \in X_{n-1}.$$

Hence if $n > m$ then

$$\|Tz_n - Tz_m\| \geq d(Tz_n, X_{n-1}) = d(y_n, X_{n-1}) = 1.$$

Consequently the bounded sequence $(z_n)_1^\infty$ does not contain a subsequence $(z_{n_k})_1^\infty$ such that Tz_{n_k} is convergent, contradicting the compactness of T. □

Our next aim is to show that if T is a compact operator and $\lambda \neq 0$ is not an eigenvalue of T then $\lambda \notin \sigma(T)$, i.e. $\lambda I - T$ is invertible. Equivalently, $\sigma(T) \cup \{0\} = \sigma_p(T) \cup \{0\}$, i.e. with the possible exception of 0, every point of the spectrum is in the point spectrum. In proving this we may and shall assume that $\lambda = 1$; so our aim is to prove that $S = I - T$ is invertible. We need two lemmas, both of which are proved in a more general form than necessary.

Lemma 5. Let $T \in \mathcal{B}_0(X)$ and set $S = I - T$. Then SX is a closed subspace of X.

Proof. Set $N = \operatorname{Ker} S$; by Theorem 4 we know that N is finite-dimensional, say with basis $\{b_1, \ldots, b_n\}$. Then there is a closed subspace $M \subset X$ such that X is the direct sum of M and N: $X = M \oplus N$. Indeed, if we choose $f_1, \ldots, f_n \in X^*$ such that $f_i(b_j) = \delta_{ij}$ then we may take $M = \bigcap_{i=1}^n \operatorname{Ker} f_i$. The projections of $x \in X$ into N and M are $p_N(x) = \sum_{i=1}^n f_i(x) b_i$ and $p_M(x) = x - p_N(s)$.

Let S_0 be the restriction of S to M: $S_0 = S|M$. Then $SX = SM = S_0M$ and $\operatorname{Ker} S_0 = \operatorname{Ker} S \cap M = \{0\}$, and so S_0 is injective. Hence to prove that $SX = S_0 M$ is closed, it suffices (and, indeed is necessary) to prove that S_0 is bounded below.

Suppose that S_0 is not bounded below, i.e. $S_0 x_n \to 0$ for some sequence $(x_n)_1^\infty \subset M$ ($\|x_n\| = 1$). Since T is compact, $(Tx_n)_1^\infty$ has a convergent subsequence, and so we may assume that $(Tx_n)_1^\infty$ itself is convergent, say $Tx_n \to y$. Then

$$x_n = (S_0 + T)x_n = S_0 x_n + Tx_n \to y,$$

and so $\|y\| = 1$. But we also have $S_0 x_n \to Sy$, and so $S_0 y = 0$, contradicting $\operatorname{Ker} S_0 = \{0\}$.   □

Lemma 6. Let $S, T \in \mathcal{B}(X)$ be such that $S + T = I$ and $SX \subset Y$, where Y is a closed proper subspace of X. Then for every $\epsilon > 0$ there is a point $x_0 \in B_X$ such that $d(Tx_0, TY) > 1 - \epsilon$.   □

Proof. By Theorem 4.8(a), there exists an $x_0 \in B_X$ such that $d(x_0, Y) > 1 - \epsilon$. As $Tx_0 = x_0 - Sx_0$, $Sx_0 \in Y$ and $TY = (I - S)Y \subset Y$, we have

$$d(Tx_0, TY) \geqslant d(x_0 - Sx_0, Y) = d(x_0, Y) > 1 - \epsilon.$$ □

Theorem 7. Let T be a compact operator and suppose $\lambda \neq 0$ is not an eigenvalue of T. Then $\lambda \notin \sigma(T)$.

Proof. By replacing T by T/λ, it suffices to prove the result for $\lambda = 1$. Let then $T \in \mathcal{B}_0(X)$, $S = I - T$ and $\operatorname{Ker} S = (0)$. We have to show that S is invertible.

Let us prove that $SX = X$. Set $Y_n = S^n X$ ($n = 0, 1, \ldots$), so that $Y_0 = X \supset Y_1 \supset \cdots$. By Lemma 5 the subspaces Y_n are closed. Let us show first that $Y_n = Y_{n+1}$ for some n. Indeed, otherwise $Y_0 \supset Y_1 \supset \cdots$ and all the inclusions are strict. Then, by Lemma 6, one can find elements $y_n \in B_{Y_n}$ such that $d(Ty_n, TY_{n+1}) > \frac{1}{2}$. But then, in particular, $\|Ty_n - Ty_m\| > \frac{1}{2}$ if $n \neq m$, and so (Ty_n) has no convergent subsequence, contradicting the compactness of T.

We claim that, in fact, $Y_0 = Y_1$. Suppose that this is not so. Then there is an m such that $Y_{m-1} \neq Y_m = Y_{m+1}$. Let $u \in Y_{m-1} \setminus Y_m$. Then, as $Su \in Y_m = Y_{m+1} = SY_m$, there exists a point $v \in Y_m$ such that $Su = Sv$. But then $S(u - v) = 0$ and so $0 \neq u - v \in \operatorname{Ker} S$, contradicting our assumption. Consequently $Y_1 = Y_0$, i.e. $SX = X$, as claimed.

The proof is essentially complete. The bounded map $S : X \to X$ is a 1–1 map of the Banach space X onto itself and so, by the inverse-mapping theorem, S is invertible.   □

Let us restate the information contained in Theorems 4 and 7 about the spectrum of a compact operator as a single result.

Theorem 8. Let T be a compact operator on an infinite-dimensional Banach space. Then $\sigma(T) = \{0, \lambda_1, \lambda_2, \ldots\}$, where the sequence $\lambda_1, \lambda_2, \ldots$ (of non-zero complex numbers) is either finite or tends to 0; furthermore, every λ_i is an eigenvalue of T, with finite-dimensional eigenspace. $\qquad\square$

With some more work, we can get more detailed information about the structure of compact operators. If T is compact and $S = I - T$ then $\operatorname{Im} S$ is a finite-codimensional closed subspace of X; even more, for a suitable $n \geqslant 1$, X is the direct sum of $\operatorname{Ker} S^n$ and $\operatorname{Im} S^n$. We prove this, and a little more, in the following theorem.

Theorem 9. Let X be a Banach space, $T \in \mathcal{B}_0(X)$ and $S = I - T$. Set $N_k = \operatorname{Ker} S^k$ and $M_k = \operatorname{Im} S^k$ ($k = 0, 1, \ldots$), where $S^0 = I$. Then $(N_k)_0^\infty$ is an increasing nested sequence of finite-dimensional subspaces and $(M_k)_0^\infty$ is a decreasing nested sequence of finite-codimensional subspaces. There is a smallest $n \geqslant 0$ such that $N_n = N_m$ for all $m \geqslant n$. Furthermore, $M_n = M_m$ for all $m \geqslant n$, X is the direct sum of M_n and N_n, and $S|M_n$ is an automorphism of M_n.

Proof. By expanding $S^k = (I - T)^k$, we see that $S^k = I - T_k$, where T_k is compact. Hence, by Lemma 5, M_k is a closed subspace of X. Clearly $N_0 \subset N_1 \subset \cdots$ and $M_0 \supset M_1 \supset \cdots$; furthermore, we know that each N_k is finite-dimensional.

As in the proof of Theorem 7, Lemma 6 implies that there is a smallest n such that $N_n = N_{n+1}$, and there is also a smallest m such that $M_m = M_{m+1}$. Then $N_n = N_{n'}$ for all $n' \geqslant n$ and $M_m = M_{m'}$ for all $m' \geqslant m$.

Let us turn to the main assertions of the theorem. We prove first that $N_n \cap M_n = \{0\}$. Let $y \in N_n \cap M_n$. As $y \in M_n$, we have $y = S^n x$ for some $x \in X$. But as $y \in N_n$, $S^n y = 0$ and so $S^{2n} x = 0$. Hence $x \in N_{2n} = N_n$, implying $S^n x = 0$. Thus $y = S^n x = 0$, showing that $N_n \cap M_n = \{0\}$.

We claim that for $p = \max\{n, m\}$ we have $X = N_n \oplus M_p$. Indeed, given $x \in X$, we have $S^p x \in M_p$. But $S^p M_p = M_p$ and so there is a vector $y \in M_p$ such that $S^p y = S^p x$. Hence $x - y \in N_p = N_n$ and so $x = y + (x - y)$ shows that $X = N_n + M_p$. Could we have $p > n$? Clearly not, since then M_n would strictly contain M_p and so we would have $N_n \cap M_n \neq \{0\}$. Thus $p = n$ and X is the direct sum of N_n and M_n, as N_n is finite-dimensional (see Exercise 4.20).

Finally, $SM_n = M_{n+1} = M_n$ and

$$\mathrm{Ker}(S|M_n) = \mathrm{Ker}\,S \cap M_n = N_1 \cap M_n \subset N_n \cap M_n = \{0\}.$$

Hence, by the inverse-mapping theorem, the restriction of S to M_n is invertible.   □

Putting Theorems 8 and 9 together, we arrive at the crowning achievement of this chapter: a rather precise description of the action of a compact operator.

Theorem 10. Let X be an infinite-dimensional Banach space and let T be a compact linear operator on X. Then $\sigma(T) = \{0, \lambda_1, \lambda_2, \ldots\}$, where the sequence $\lambda_1, \lambda_2, \ldots$ is either finite or tends to 0. For every $\lambda = \lambda_i$ there is an integer $k_\lambda \geq 1$ and closed subspaces $N_\lambda = N(\lambda; T)$ and $M_\lambda = M(\lambda; T)$ invariant under T, such that N_λ is finite-dimensional and $X = N_\lambda \oplus M_\lambda$.

The restriction of $\lambda I - T$ to M_λ is an automorphism of M_λ,

$$N_\lambda = \mathrm{Ker}(\lambda I - T)^{k_\lambda} \neq \mathrm{Ker}(\lambda I - T)^{k_\lambda - 1},$$

and for $\mu = \lambda_j \neq \lambda = \lambda_i$ we have $N_\lambda \subset M_\mu$.

Proof. Only the last claim needs justification. The operator $\mu I - T$ maps M_λ into itself and N_λ into itself. Furthermore, $(\mu I - T)|N_\lambda$ is an automorphism of the finite-dimensional space N_λ since if we had $(\mu I - T)x = 0$ for some $x \in N_\lambda$ ($x \neq 0$) then we would have $(\lambda I - T)^n x = (\lambda - \mu)^n x \neq 0$ for every $n \geq 1$, contradicting $(\lambda I - T)^{k_\lambda} x = 0$. Consequently, $(\mu I - T)^n N_\lambda = N_\lambda$ for every $n \geq 1$, and so $N_\lambda \subset M_\mu$.   □

There is no doubt that Theorem 10 is a very beautiful theorem. At first sight it is not only beautiful but very impressive as well: it seems to come close to giving us a very fine decomposition of the space into a direct sum of generalized eigenspaces. Unfortunately, this is rather a mirage: the theorem cannot even guarantee that our compact operator has a non-trivial closed invariant subspace, let alone give a direct-sum decomposition. In fact, non-trivial closed invariant subspaces *do* exist, as we shall prove in Chapter 16. However, to prove the existence of invariant subspaces we shall need some results to be proved in Chapter 15. Before we turn to that, in the next chapter we shall show that a best possible decomposition *can* be guaranteed if we deal with a compact normal operator on a Hilbert space.

Exercises

1. Let $1 \leqslant p < \infty$, $a = (a_i)_1^\infty \in l_p$ and

$$A = \{x = (x_i)_1^\infty \in l_p : |x_i| \leqslant |a_i| \ (i = 1, 2, \ldots)\}.$$

   Prove that A is a compact subset of l_p.
2. Let K be a closed and bounded subset of l_p ($1 \leqslant p < \infty$). Prove that K is compact if and only if for every $\epsilon > 0$ there is an n such that $\sum_{i=n}^\infty |x_i|^p < \epsilon$ for every $x = (x_i)_1^\infty \in K$.
3. As in Example 1(xiii) of Chapter 2, let $C^{(k)}(0,1)$ be the vector space of k times continuously differentiable functions $f : (0,1) \to \mathbb{R}$ such that

$$\|f\|_k = \sup \sum_{l=0}^k |f^{(l)}(t)| < \infty.$$

   Show that $X_k = (C^{(k)}(0,1), \|\cdot\|_k)$ is a Banach space and the formal identity map $i : X_k \to X_{k-1}$ ($f \mapsto f$) is a compact operator.
4. Let $T \in \mathcal{B}_0(X, Y)$, where X is infinite-dimensional. Show that the closure of $TS(X) = \{Tx : x \in X, \|x\| = 1\}$ in Y contains 0. (HINT: Consider a sequence $(x_n) \subset S(X)$ such that $\|x_n - x_m\| \geqslant 1$ for $n \neq m$.)
5. Let $(e_n)_1^\infty$ be an orthonormal basis of a Hilbert space H and let $T \in \mathcal{B}_0(H, Y)$, where Y is a normed space. Show that $Te_n \to 0$.
6. Let X be a normed space such that for every finite set $A \subset X$ and every $\epsilon > 0$, X has a decomposition

$$X = M \oplus N$$

   as a direct sum of two closed subspaces, such that M is finite-dimensional,

$$d(a, M) < \epsilon$$

   for every $a \in A$, and

$$\|p_N(x)\| \leqslant c d(x, M)$$

   for some $c > 0$ and every $x \in X$, where p_N is the canonical projection onto N. Show that $\mathcal{B}_0(X)$ is the norm closure of $\mathcal{B}_{00}(X)$, i.e.

every compact operator on X is the operator-norm limit of finite rank operators.

7. Let X be a Banach space with a Schauder basis $(x_n)_1^\infty$. Show that $\mathcal{B}_0(X)$ is the closure of $\mathcal{B}_{00}(X)$.

8. Let $(\lambda_n)_1^\infty$ be a sequence of non-zero complex numbers tending to 0. Show that $\sigma(T) = \{0,\lambda_1,\lambda_2,\ldots\}$ for some complex operator T on some compact Banach space X. Show that if all λ_i are real then X can be chosen to be a real Banach space.

9. Let X and Y be Banach spaces, let $T \in \mathcal{B}_0(X,Y)$, and let $J \in \mathcal{B}(X,Y)$ be invertible. Show that $\mathrm{Im}(J-T)$ is closed in Y and has finite codimension.

10. Let X be a complex Banach space, and let $T \in \mathcal{B}(X)$ be such that T^n is compact for some $n \geq 1$. Show that $\sigma(T) = \{0,\lambda_1,\lambda_2,\ldots\}$, where the sequence $\lambda_1,\lambda_2,\ldots$ is finite or tends to 0, and every λ_i is an eigenvalue of T. What is the relationship between the subspaces $N(\lambda;T)$ and $N(\mu;T^n)$?

11. Let $X_1,X_2,\ldots$ be Banach spaces and let $X = \bigoplus_{n=1}^\infty X_n$ be the direct sum of these spaces, with $x = (x_n)_1^\infty$ having norm

$$\|\|x\|\| = \sum_{n=1}^\infty \|x_n\|.$$

Let $T_n \in \mathcal{B}(X_n)$ $(n = 1,2,\ldots)$ and set $T = \bigoplus_{n=1}^\infty T_n$, that is, define $T \in \mathcal{B}(X)$ by $Tx = (Tx_n)_1^\infty$. Show that $T \in \mathcal{B}_0(X)$ iff $T_n \in \mathcal{B}_0(X_n)$ for every n.

12. Let X be a Banach space, $(T_n)_1^\infty \subset \mathcal{B}_0(X)$, and let $T_n \to T$ in the operator-norm topology. Show that $\bigcup_{n=1}^\infty T_n(B)$ is relatively compact, where $B = B(X)$ is the unit ball of X. Show also that if $\lambda_n \in \sigma(T_n)$ and $\lambda_n \to \lambda$ then $\lambda \in \sigma(T)$.

13. Let E be the space $C[0,1]$, endowed with the *Euclidean norm*

$$\|f\|_2 = \left(\int_0^1 |f(x)|^2 \, dx\right)^{1/2},$$

and let T be as in Example 1 (iii). Show that $T \in \mathcal{B}_0(E)$, i.e. T maps the unit ball of the Euclidean space E into a relatively compact set.

14. Let H be a Hilbert space and $T \in \mathcal{B}(H)$. Show that $T \in \mathcal{B}_0(H)$ if and only if $Tx_n \to 0$ whenever x_n converges weakly to 0, i.e. whenever $(x_n,x) \to 0$ for every $x \in H$.

15. Construct a compact operator T on l_p $(1 \leq p \leq \infty)$ such that $\sigma(T) = \{0\}$ and 0 is not an eigenvalue of T.

16. Let $c_1, c_2, \ldots$ be non-negative reals and

$$C = \{x \in l_2 : x = (x_k)_1^\infty, \ |x_k| \leq c_k \text{ for every } k\}.$$

Show that if C is a compact subset of l_2 then $c_k \to 0$. For what sequences $(c_k)_1^\infty$ is C compact?

17^+. Let K be a compact subset of a normed space. Show that K is contained in the closed absolutely convex hull of a sequence tending to 0: there is a sequence $x_n \to 0$ such that $K \subset \bar{C}$, where

$$C = \mathrm{co}\{x_n : n = 1, 2, \ldots\} = \left\{ \sum_{i=1}^{n} \lambda_i x_i : \sum_{i=1}^{n} |\lambda_i| \leq 1, \ n = 1, 2, \ldots \right\}.$$

Notes

Compact operators were first introduced and applied by Hilbert, *Grundzüge einer allgemeinen Theorie der linearen Integralgleichungen*, Leipzig, 1912, and F. Riesz, *Les systèmes d'équations à une infinité d'inconnus*, Paris, 1913, and *Über lineare Funktionalgleichungen*, Acta Math., **41** (1918), 71–98. In presenting the Riesz theory of compact operators we relied on Ch. XI of J. Dieudonné, *Foundations of Modern Analysis*, Academic Press, New York and London, 1960, xiv + 361 pp.

Theorem 3 is due to J. Schauder, *Über lineare vollstetige funktional Operationen*, Studia Math., **2** (1930), 185–96.

The first solution of the approximation problem was published by P. Enflo, *A counterexample to the approximation problem in Banach spaces*, Acta Math., **30** (1973), 309–17; a simplified version of the solution is in A. M. Davie, *The approximation problem for Banach spaces*, Bull. London Math. Soc., **5** (1973), 261–6.

14. COMPACT NORMAL OPERATORS

In the previous chapter we saw that for every compact operator T on a Banach space X, the space can *almost* be written as a direct sum of generalized eigenspaces of T. If we assume that X is not merely a Banach space, but a Hilbert space, and T is not only compact but compact and normal, then such a decomposition is indeed possible – in fact, there is a decomposition with even better properties. Such a decomposition will be provided by the *spectral theorem for compact normal operators*: a complete and very simple description of compact normal operators. Thus with the study of a compact normal operator on a Hilbert space we arrive in the promised land: everything fits, everything works out beautifully, there are no blemishes. This is the best of all possible worlds.

We shall give two proofs of the spectral theorem, claiming the existence of the desired decomposition. In the first proof we shall make use of some substantial results from previous chapters, including one of the important results concerning the spectrum of a compact operator. The second proof is self-contained: we shall replace the results of the earlier chapters by easier direct arguments concerning Hilbert spaces and normal operators.

To start with, we collect a number of basic facts concerning normal operators in the following lemma. Most of these facts have already been proved, but for the sake of convenience we prove them again.

Lemma 1. Let $T \in \mathcal{B}(H)$ be a normal operator. Then the following assertions hold.
 (a) $\|Tx\| = \|T^*x\|$ for every $x \in H$.
 (b) $\operatorname{Ker} T = \operatorname{Ker} T^*$.
 (c) $\|T^n\| = \|T\|^n$ for every $n \geq 1$.

(d) $r(T) = \|T\|$.

(e) If $\lambda \neq \mu$ then $\mathrm{Ker}(\lambda I - T) \perp \mathrm{Ker}(\mu I - T)$.

(f) For every $\lambda \in \mathbb{C}$, both $\mathrm{Ker}(\lambda I - T)$ and $(\mathrm{Ker}(\lambda I - T))^{\perp}$ are invariant under both T and T^*.

(g) If H is the orthogonal direct sum of the closed subspaces H_0 and H_1 invariant under T then with $T_0 = T|H_0$ and $T_1 = T|H_1$ we have

$$\|T\| = \max\{\|T_0\|, \|T_1\|\}$$

and T_i is a normal operator on H_i, with $T_i^* = T^*|H_i$ $(i = 1, 2)$.

Proof. (a) As $T^*T = TT^*$, we have

$$\|Tx\|^2 = (Tx, Tx) = (T^*Tx, x) = (TT^*x, x) = (T^*x, T^*x) = \|T^*x\|^2.$$

(b) By part (a), we have $Tx = 0$ iff $T^*x = 0$.

(c) If $S \in \mathscr{B}(H)$ is hermitian then

$$\|Sx\|^2 = (Sx, Sx) = (S^*Sx, x) = (S^2x, x) \leqslant \|S^2\|\|x\|^2.$$

From this it follows that $\|S^2\| = \|S\|^2$, and by induction on m we get $\|S^{2^m}\| = \|S\|^{2^m}$. This implies that $\|S^n\| = \|S\|^n$ for every $n \geqslant 1$. As TT^* is hermitian,

$$\|T^n\|^2 = \|T^n(T^n)^*\| = \|(TT^*)^n\| = \|TT^*\|^n = \|T\|^{2n}.$$

(d) By (c) and the spectral-radius formula (Theorem 12.9),

$$r(T) = \lim_{n\to\infty} \|T^n\|^{1/n} = \lim_{n\to\infty} \|T\| = \|T\|.$$

(In fact, (c) and (d) are equivalent: if $S \in \mathscr{B}(X)$ for a complex Banach space X then $r(S) = \|S\|$ iff $\|S^n\| = \|S\|^n$ for every $n \geqslant 1$.)

(e) If $Tx = \lambda x$ and $Ty = \mu y$ then $T^*y = \bar{\mu}y$ because by (b) we have $y \in \mathrm{Ker}(\mu I - T) = \mathrm{Ker}(\bar{\mu}I - T^*)$. Therefore

$$\lambda(x, y) = (Tx, y) = (x, T^*y) = (x, \bar{\mu}y) = \mu(x, y),$$

and so if $\lambda \neq \mu$ then $(x, y) = 0$.

(f) As $\lambda I - T$ commutes with T and T^*, $\mathrm{Ker}(\lambda I - T)$ is invariant under both T and T^*.

Also, let $(x, y) = 0$ for all $y \in \mathrm{Ker}(\lambda I - T)$. Then, since $\mathrm{Ker}(\lambda I - T)$ is invariant under T^*, for $y \in \mathrm{Ker}(\lambda I - T)$ we have

$$(Tx, y) = (x, T^*y) = 0.$$

Hence $Tx \in (\mathrm{Ker}(\lambda I - T))^{\perp}$. Similarly,

$$(T^*x, y) = (x, Ty) = 0$$

for every $y \in \text{Ker}(\lambda I - T)$ and so $T^*x = (\text{Ker}(\lambda I - T))^\perp$.

(h) Let $x = h_0 + h_1$, with $h_i \in H_i$ $(i = 0, 1)$. Then

$$\|x\|^2 = \|h_0\|^2 + \|h_1\|^2, \qquad Tx = Th_0 + Th_1 = T_0 h_0 + T h_1$$

and

$$\begin{aligned}
\|Tx\|^2 &= \|T_0 h_0\|^2 + \|T_1 h_1\|^2 \\
&\leq \|T_0\|^2 \|h_0\|^2 + \|T_1\|^2 \|h_1\|^2 \\
&\leq \max\{\|T_0\|^2, \|T_1\|^2\}(\|h_0\|^2 + \|h_1\|^2) \\
&= \max\{\|T_0\|^2, \|T_1\|^2\} \|x\|^2.
\end{aligned}$$

Thus $\|T\| \leq \max\{\|T_1\|, \|T_2\|\}$. The reverse inequality is obvious.

Finally, as H_0 and H_1 are invariant under T, it follows that $H_1 = H_0^\perp$ and $H_0 = H^\perp$ are invariant under T^*. $\qquad \square$

It is worth emphasizing that Lemma 1 is a collection of elementary and simple facts, *except* for part (d), which is based on the spectral-radius formula.

Let us see then the first incarnation of the spectral theorem, claiming the existence of a *spectral decomposition* for a compact normal operator.

Theorem 2. Let $T \in \mathcal{B}(H)$ be a compact normal operator. For an eigenvalue λ of T, let $H_\lambda = \text{Ker}(T - \lambda I)$ be the eigenspace of T belonging to λ, and denote by P_λ the orthogonal projection onto H_λ. The operator T has countably many non-zero eigenvalues, say $\lambda_1, \lambda_2, \dots$. Furthermore, $\dim H_{\lambda_k} < \infty$ for every k, the projections P_{λ_l} are orthogonal, i.e. $P_{\lambda_k} P_{\lambda_l} = 0$ if $k \neq l$, and

$$T = \sum_k \lambda_k P_{\lambda_k}, \tag{1}$$

where the series is convergent in the norm of $\mathcal{B}(H)$.

Proof. By Theorem 13.8 and Lemma 1 (e), we have to prove only (1).

Given $\epsilon > 0$, choose $n \geq 1$ such that $|\lambda_k| < \epsilon$ for $k > n$. Set

$$H_0 = \sum_{k=1}^n H_{\lambda_k}, \qquad H_1 = H_0^\perp, \qquad S = \sum_{k=1}^n \lambda_k P_{\lambda_k}.$$

Then H_0 and H_1 are invariant under S and T. With $T_i = T|H_i$ and $S_i = S|H_i$ $(i = 1, 2)$, we have $T_0 = S_0$ and $S_1 = 0$. Therefore, by

Lemma 1 (g),

$$\|T - S\| = \max\{\|T_0 - S_0\|, \|T_1 - S_1\|\} = \|T_1\|.$$

But T_1 is a compact normal operator and so, by Theorem 13.8, $\|T_1\|$ is precisely the maximum modulus of an eigenvalue of T_1. As every eigenvalue of T_1 is an eigenvalue of T, by our choice of n we have $\|T_1\| < \epsilon$. Hence (1) does hold. $\qquad\square$

Let us state two other versions of the spectral theorem.

Theorem 3. Let T be a compact hermitian operator on an infinite-dimensional Hilbert space H. Then one can find a closed subspace H_0 of H, a (finite or countably infinite) orthonormal basis (x_n) of H_0, and a sequence of complex numbers $\nu_n \to 0$, such that if $x = \sum_n c_n x_n + x'$, where $x' \in H_0^\perp$, then

$$Tx = \sum_n \nu_n c_n x_n.$$

Proof. Let $\lambda_1, \lambda_2, \ldots$ and $H_{\lambda_1}, H_{\lambda_2}, \ldots$ be as in Theorem 2. Take a (necessarily finite) orthonormal basis in each H_{λ_k} and let (x_n) be the union of these bases. Let H_0 be the closed linear span of the orthonormal sequence (x_n), and set $\nu_n = \lambda_k$ if $x_n \in H_{\lambda_k}$. $\qquad\square$

Corollary 4. Let T be a compact normal operator on a Hilbert space H. Then H has an orthonormal basis consisting of eigenvalues of T. $\qquad\square$

In fact, compact normal operators are characterized by Theorem 2 (or Theorem 3). Let $\{x_\gamma : \gamma \in \Gamma\}$ be an orthonormal basis of a Hilbert space H, and let $T \in \mathcal{B}(H)$ be such that $Tx_\gamma = \nu_\gamma x_\gamma$. Then T is compact iff

$$|\{\gamma : |\nu_\gamma| \geq \epsilon, \gamma \in \Gamma\}| < \infty \qquad (2)$$

for every $\epsilon > 0$ (see Exercise 2).

Our proof of Theorem 2 was based on two substantial results: Theorem 13.8 concerning compact operators on Banach spaces, and the spectral-radius formula. We shall show now how one can prove Theorem 2 without relying on these results. It is a little more convenient to prove Theorem 2 for compact *hermitian* operators; it is then a simple matter to extend it to normal operators.

Recall that the *numerical range* $V(T)$ of a Hilbert space operator $T \in \mathcal{B}(H)$ is

$$\{(Tx,x) : x \in S(H)\}$$

and the *numerical radius* $v(T)$ is

$$v(T) = \sup\{|\lambda| : \lambda \in V(T)\}.$$

If $T \in \mathcal{B}(H)$ is hermitian, i.e. $T^* = T$, then its numerical range is real since

$$(Tx,x) = (x, T^*x) = (x, Tx) = \overline{(Tx,x)}$$

for every $x \in H$ and so (Tx,x) is real. In fact, $T \in \mathcal{B}(H)$ is hermitian iff its numerical range is real. Also, the spectrum of a hermitian operator is real. We shall not make use of any of the results proved about numerical ranges; the next lemma is proved from first principlues.

Lemma 5. Let T be a hermitian operator. Then $\|T\| = v(T)$.

Proof. Set $v = v(T)$, so that $|(Tx,x)| \leqslant v\|x\|^2$ for every $x \in H$. We have to show that $\|T\| \leqslant v$.

Given $x \in S(H)$, let $y \in S(H)$ be such that $Tx = \|Tx\|y$. Then $(Tx,y) = (x, Ty) = \|Tx\|$ and so

$$\begin{aligned}
\|Tx\| = (Tx,y) &= \tfrac{1}{4}\{(T(x+y),(x+y)) - (T(x-y),(x-y))\} \\
&\leqslant \tfrac{1}{4}v\{\|x+y\|^2 + \|x-y\|^2\} \\
&= \tfrac{1}{2}v\{\|x\|^2 + \|y\|^2\} = v.
\end{aligned}$$

Hence $\|Tx\| \leqslant v$ for every $x \in S(H)$ and so $\|T\| \leqslant v$, as claimed. □

Theorem 6. Let U be a compact hermitian operator on H. Then U has an eigenvalue of absolute value $\|U\|$.

Proof. Set

$$a = \inf_{\|x\|=1} (Ux,x) \qquad \text{and} \qquad b = \sup_{\|x\|=1} (Ux,x)$$

so that $\overline{V(U)} = [a,b]$. By Lemma 5, $\|U\| = \max\{-a,b\}$. Replacing U by $-U$, if necessary, we may assume that $\|U\| = b > 0$. We have to show that b is an eigenvalue of U.

By the definition of b, there is a sequence $(x_n) \subset S(H)$ such that $(Ux_n, x_n) \to b$. Since U is a compact operator, by replacing (x_n) by a subsequence, we may suppose that (Ux_n) is convergent, say $Ux_n \to y_0$. Then $\|y_0\| \geqslant b$ because $(Ux_n, x_n) \to b$ and $\|x_n\| = 1$. As

$$\|Ux_n - bx_n\|^2 = \|Ux_n\|^2 - 2b(Ux_n, x_n) + b^2\|x_n\|^2$$
$$\leq \|U\|^2 - 2b(Ux_n, x_n) + b^2$$
$$= 2b^2 - 2b(Ux_n, x_n).sp - .05$$

and

$$(Ux_n, x_n) \to b,$$

we have

$$(U-b)x_n \to 0.$$

Therefore

$$x_n = \frac{1}{b}\{(U-(U-b))x_n\} \to \frac{y_0}{b}.$$

Put $x_0 = y_0/b$. Then, on the one hand, $Ux_n \to y_0 = bx_0$ and, on the other hand, $Ux_n \to Ux_0$. Consequently we have $Ux_0 = bx_0$. Since $\|x_0\| \geq 1$ (in fact, $\|x_0\| = 1$), b is indeed an eigenvalue of U. $\square$

Let us now see how Theorem 6 may be used to deduce Theorem 2 for compact hermitian operators. For the sake of variety, we restate Theorem 2 in the following form.

Theorem 7. Let H be a Hilbert space and let $U \in \mathcal{B}(H)$ be a compact hermitian operator. Then there is a (possibly finite) sequence (λ_k) of real numbers and a sequence (H_k) of linear subspaces of H such that
   (a) $\lambda_k \to 0$;
   (b) $\dim H_k < \infty$;
   (c) $H_k \perp H_l$ for $k \neq l$,
   (d) if $x = \sum_k x_k + \tilde{x}$, where $x_k \in H_k$ and $\tilde{x} \in H_k^\perp$ for every k, then

$$Ux = \sum_k \lambda_k x_k.$$

Proof. Let λ_γ, $(\gamma \in \Gamma)$ be the non-zero eigenvalues of U and let H_γ be the eigenspace belonging to λ_γ: $H_\gamma = \mathrm{Ker}(U - \lambda_\gamma I)$. We know that $H_\gamma \perp H_\delta$ if $\gamma \neq \delta$.

Let us show first that $\dim H_\gamma < \infty$ and, for every $\epsilon > 0$, there are only finitely many λ_γ with $|\lambda_\gamma| \geq \epsilon$. Suppose not. Then, by taking an orthonormal basis in each H_γ with $|\lambda_\gamma| \geq \epsilon$, we find that there is an infinite orthonormal sequence $(x_n)_1^\infty$ such that $Ux_n = \mu_n x_n$, where $|\mu_n| \geq \epsilon$. But then $(Ux_n)_1^\infty$ does not contain a convergent subsequence, contradicting the compactness of U.

This implies that the non-zero eigenvalues may be arranged in a sequence (λ_k) such that with $H_k = \mathrm{Ker}(U - \lambda_k I)$ the conditions (a)–(c) are satisfied.

Then, as each H_k is invariant under U, so is the closed linear span M of all the H_k and, consequently, so is $M^{\perp}$. Denote by $\tilde{U}$ the restriction of U to $M^{\perp}$. Then $\tilde{U} \in \mathcal{B}(M^{\perp})$ is also a compact hermitian operator. As a non-zero eigenvalue of $\tilde{U}$ is also a non-zero eigenvalue of U, it follows from the definition of M and from Theorem 6, that $\tilde{U} = 0$.

If the sequence (λ_k) of non-zero eigenvalues is finite then we are done. Otherwise, let

$$x = \sum_{k=1}^{\infty} x_k + \tilde{x}, \qquad \text{where } x_k \in H_k \text{ and } \tilde{x} \in M^{\perp}.$$

Put

$$x^{(n)} = \sum_{k=1}^{n} x_k + \tilde{x} \qquad \text{and} \qquad y^{(n)} = \sum_{k=1}^{n} \lambda_k x_k.$$

Then $Ux^{(n)} = y^{(n)}$ and $x^{(n)} \to x$. As

$$y^{(n)} \to y = \sum_{k=1}^{\infty} \lambda_k x_k \in H,$$

the continuity of U implies that $Ux = y$, proving (d).   □

Before we recover from Theorem 7 the full force of Theorem 2, let us show that compact hermitian operators are rather like real numbers. An operator $T \in \mathcal{B}(H)$ is said to be *positive* if it is hermitian and $V(T) \subset [0, \infty)$, i.e. $(Tx, x) \geqslant 0$ for every $x \in H$. Note that if T is *any* (bounded linear) operator on a Hilbert space then T^*T and TT^* are positive (hermitian) operators:

$$(T^*Tx, x) = (Tx, Tx) = \|Tx\|^2 \qquad \text{and} \qquad (TT^*x, x) = \|T^*x\|^2.$$

Theorem 8. A compact positive operator U on a Hilbert space has a unique positive square root V. Every hermitian square root of U is compact.

Proof. Let $\lambda_1, \lambda_2, \ldots$ be the non-zero eigenvalues of U, let H_k be the eigenspace belonging to λ_k and let M be the closed linear span of the H_k. Then $M^{\perp} = \mathrm{Ker}\, U$ and $\lambda_k > 0$ for every k. Define $V \in \mathcal{B}(H)$ by $Vx = \sqrt{\lambda_k}\, x$ if $x \in H_k$ and $Vx = 0$ if $x \in M^{\perp}$. Then V is a positive square root of U.

Now let W be a hermitian square root of U. Note that W commutes with U and so it commutes with $U - \lambda_k I$: $UW = W^2 W = WW^2 = WU$. Therefore $H_k = \text{Ker}(U - \lambda_k I)$ is invariant under W, and so is $M^\perp = \text{Ker}\, U$. Then $W^2 | M^\perp = 0$ and so $W | M^\perp = 0$.

Also, let $(x_1, \ldots, x_n)$ be an orthonormal basis of H_k consisting of eigenvectors of W: $Wx_i = \mu_i x_i$, say. As $W^2 = U$, we have $\mu_i^2 = \lambda_k$ and so $\mu_i = \pm\sqrt{\lambda_k} \to 0$. Since $\sqrt{\lambda_k} \to 0$, by relation (2) the operator W is compact. Furthermore, if W is positive then $\mu_i = \sqrt{\lambda_k}$ and so W is precisely V. $\qquad\qquad\square$

For $T \in \mathcal{B}_0(H)$ the unique positive square root $(T^*T)^{1/2}$ of the compact positive operator T^*T, guaranteed by Theorem 8, is called the *modulus* or *absolute value* of T, and is denoted by $|T|$. Note that $\||T|x\| = \|Tx\|$ for every $x \in H$, because

$$\||T|x\|^2 = \langle |T|x, |T|x \rangle = \langle (T^*T)^{1/2}x, (T^*T)^{1/2}x \rangle$$
$$= \langle T^*Tx, x \rangle = \langle Tx, Tx \rangle = \|Tx\|^2.$$

In fact, $|T|$ is the unique positive operator with this property (see Exercise 4).

Let us see then how the spectral theorem for compact normal operators can be recovered from Theorems 7 and 8. To be precise, we shall deduce the version given in Corollary 4 from these theorems.

Given a compact normal operator T, how can we get an orthonormal basis consisting of eigenvectors of T?

Let $H_1, H_2, \ldots$ be the eigenspaces belonging to the non-zero eigenvectors of the compact positive operator $U = TT^* = TT^*$, and let

$$H_0 = \text{Ker}\, U = \bigcap_{k \geq 1} H_k^\perp = (\overline{\text{lin}}\{H_1, H_2, \ldots\})^\perp.$$

Since $TU = UT$, each H_k is invariant under T. For $x \in H_0$ we have

$$\|Tx\|^2 = \langle Tx, Tx \rangle = \langle T^*Tx, x \rangle = \langle Ux, x \rangle = 0$$

and so $T|H_0 = 0$. Furthermore, for $k \geq 1$ the restriction of T to H_k is normal; as H_k is finite-dimensional, we know from linear algebra that H_k has an orthonormal basis consisting of eigenvectors of T (see Exercise 5). Take the union of these bases, together with an orthonormal basis of H_0.

It is easily seen that a similar assertion holds for several commuting compact normal operators.

Theorem 9. Let $T_1, \ldots, T_n$ be commuting compact normal operators on a Hilbert space H. Then H has an orthonormal basis consisting of common eigenvectors of all the T_i.

Proof. For every $\mu \in \mathbb{C}$ and $k = 1, \ldots, n$, the eigenspace $\operatorname{Ker}(\mu I - T_k)$ is invariant under all the T_i. Hence H is the orthogonal direct sum of the subspaces

$$H_{\mu_1, \ldots, \mu_n} = \bigcap_{i=1}^n \operatorname{Ker}(\mu_i I - T_i).$$

All these spaces are finite-dimensional, with the possible exception of $H_{0, \ldots, 0}$. Taking an orthonormal basis of each $H_{\mu_1, \ldots, \mu_n}$, the union of these bases will do. $\square$

As our final theorem concerning abstract operators in this chapter, let us note that our results, say Theorem 2 or Theorem 3, give a complete characterization of compact normal operators up to unitary equivalence. Two operators $T, T' \in \mathcal{B}(H)$ are said to be *unitarily equivalent* if for some unitary operator U we have $T' = U^{-1}TU = U^*TU$, i.e. if they have the same matrix representation with respect to some orthonormal bases.

Let $\mathcal{N}$ be the collection of functions $n : \mathbb{C} \backslash \{0\} \to \{0, 1, 2, \ldots\}$ whose support $\{\lambda \in \mathbb{C} \backslash \{0\} : n(\lambda) \geq 1\}$ has no accumulation point (i.e. in $\mathbb{C}$ there is no accumulation point other than 0). In particular, the support is finite or countably infinite. The following result is easily read out of Theorem 2 (see Exercise 14).

Theorem 10. Let H be an infinite-dimensional complex Hilbert space and let $\mathcal{C}_N(H)$ be the collection of compact normal operators on H. For $T \in \mathcal{C}_N(H)$ and $\lambda \in \mathbb{C} \backslash \{0\}$ set

$$n_T(\lambda) = \dim \operatorname{Ker}(\lambda I - T).$$

Then the correspondence $T \mapsto n_T$ defines a surjection $\mathcal{C}_N(H) \to \mathcal{N}$; furthermore, T and T' are unitarily equivalent iff $n_T = n_{T'}$. $\square$

We close this chapter by showing how the spectral theorems we have just proved enable us to solve a Fredholm integral equation.

Let $I = [a, b]$ for some $a < b$, and write E for the Euclidean space $C(I)$ endowed with the inner product

$$(f, g) = \int_a^b f(t)\overline{g(t)} \, dt$$

and norm $\|f\|_2 = (f, f)^{1/2}$. Thus the completion of E is $L_2(0, 1)$.

Let $K(s,t) \in C(I \times I)$, and for $f \in E$ define $Uf \in E$ by

$$(Uf)(s) = \int_a^b K(s,t)f(t) \, dt.$$

Then K is the *kernel* of the integral operator U. It is easily checked that $U \in \mathscr{B}(E, C(I))$ and U maps the unit ball of E into a relatively compact set in $C(I)$. Indeed, this follows from the Arzelá–Ascoli theorem, since if $|K(s,t) - K(s',t)| < \epsilon$ for all t then $|(Uf)(s) - (Uf)(s')| \leq \epsilon \|f\|_2$ by the Cauchy–Schwarz inequality.

As the formal identity map $C(I) \to E$, where $f \mapsto f$, is continuous, U extends to a compact operator on $L_2(0,1)$; for simplicity, we write U for this extension as well. In fact,

$$(Uf)(s) = \int_a^b K(s,t)f(t) \, dt$$

for $f \in L_2(0,1)$.

From now on we suppose also that $K(s,t) = \overline{K(t,s)}$; in this case U is easily seen to be a hermitian operator. We consider a *Fredholm integral equation*

$$g(s) = \int_a^b K(s,t)f(t) \, dt - \lambda f(s) = ((U-\lambda)f)(s).$$

For what values of λ can we solve this equation, and what can we say about the solution? As we shall see, this question is ideally suited for the theory we have at our disposal.

Let us prove three quick lemmas before giving the answer.

Lemma 11. Let (φ_n) be an orthonormal sequence of eigenvectors of U with non-zero eigenvalues (λ_n) guaranteed by each of Theorems 2, 3 and 7, and Corollary 4, such that if $f \perp \varphi_n$ for every n then $Uf = 0$. (The sequences (φ_n) and (λ_n) may thus be finite or infinite). Then each λ_n is real,

$$\sum_n |\lambda_n \varphi_n(t)|^2 \leq N$$

for every x, where N depends only on $K(s,t)$, and

$$\sum_n \lambda_n^2 \leq \int_a^b \int_a^b |K(s,t)|^2 \, ds \, dt.$$

Proof. For $0 \leq t \leq 1$ put $k_t(s) = K(s,t)$. Then, by Bessel's inequality, for every m we have

$$\sum_{n=1}^{m} \left| \int_{a}^{b} k_t(s)\,\varphi_n(s)\ ds \right|^2 = \sum_{n=1}^{m} |\lambda_n \varphi_n(t)|^2 \leqslant \int_{a}^{b} |k_t(s)|^2\ ds.$$

Therefore

$$\sum_{n=1}^{m} \lambda_n^2 = \int_{a}^{b} \sum_{n=1}^{m} |\lambda_n \varphi_n(t)|^2\ dx \leqslant \int_{a}^{b} \int_{a}^{b} |K(s,t)|^2\ ds\ dt. \qquad \Box$$

Lemma 12. If $\varphi_n \in L_2(0,1)$ is an eigenvector of U with eigenvalue $\lambda_n \neq 0$ then $\varphi_n \in C(I)$.

Proof. By the Cauchy–Schwarz inequality, $Uf \in C(I)$ for $f \in L_2(0,1)$. Hence $U\varphi_n = \lambda_n \varphi_n \in C(I)$. $\qquad \Box$

Lemma 13. Let $g \in L_2(0,1)$, $f = Ug$ and let (c_n) be the Fourier coefficients of f with respect to (φ_n). Then the series $\sum_n c_n \varphi_n(s)$ converges absolutely and uniformly to $f(s)$ in $[0,1]$.

Proof. Let $g = \sum_n a_n \varphi_n + \tilde{g}$ be the decomposition of g in $L_2(0,1)$, where $\tilde{g} \perp \varphi_n$ for every n. As $U \colon L_2(0,1) \to C(I)$ is continuous, $c_n = \lambda_n a_n$ and $U\tilde{g} = 0$, the uniform convergence follows. Furthermore, for every finite set F of indices,

$$\left(\sum_{n \in F} |c_n \varphi_n(s)| \right)^2 = \left(\sum_{n \in F} |a_n \lambda_n \varphi_n(s)| \right)^2$$

$$\leqslant \left(\sum_{n \in F} |a_n|^2 \right)\left(\sum_{n \in F} |\lambda_n \varphi_n(s)|^2 \right) \leqslant N\|g\|_2^2$$

by Lemma 11. $\qquad \Box$

We are now ready to answer our question about the Fredholm integral equation.

Theorem 14. If $\lambda \notin \sigma(U)$ and $g \in C(I)$ then the Fredholm integral equation

$$Uf - \lambda f = g$$

with a hermitian kernel $K(s,t) = \overline{K(t,s)}$ has a unique solution f given by

$$f(t) = -\frac{1}{\lambda}g(t) + \sum_n \frac{\lambda_n}{\lambda(\lambda_n - \lambda)} a_n \varphi_n(t),$$

where (φ_n) is an orthonormal basis of eigenvectors of U, the series is absolutely and uniformly convergent in $[0,1]$ and the (a_n) are the Fourier coefficients of g with respect to (φ_n), i.e.

$$a_n = \int_a^b g(t)\overline{\varphi_n(t)}\, dt.$$

Proof. The unique solution in $L_2(0,1)$ is $f = (U-\lambda)^{-1}g$. Clearly, f has Fourier coefficients $a_n/(\lambda_n-\lambda)$. By Lemma 13, $g+\lambda f = Uf \in C(I)$ and

$$g+\lambda f = \sum_n \frac{\lambda_n a_n}{\lambda_n-\lambda}\varphi_n,$$

where the series is absolutely and uniformly convergent. $\qquad\square$

There are numerous other applications of the elementary spectral theory of compact hermitian operators to integral equations, notably to the Sturm–Liouville equation. However, a proper account of these applications would be much longer than we have space for.

In conclusion, let us point out that the spectral theorem for compact normal operators is only the beginning of the story. The result we proved is a very simple version of a spectral theorem for normal (not necessarily compact) operators. With every normal operator one can associate a so-called *spectral measure* containing all the information about the operator up to unitary equivalence.

Exercises

1. Let H_0 be a closed subspace of a Hilbert space H invariant under a normal operator $T \in \mathcal{B}(H)$. Is $H_1 = H_0^\perp$ necessarily invariant under T?

2. Prove relation (2), i.e. show that if $\{x_\gamma : \gamma \in \Gamma\}$ is an orthonormal basis of a Hilbert space H and $T \in \mathcal{B}(H)$ is given by $Tx_\gamma = \nu_\gamma x_\gamma$, where $\nu_\gamma \in \mathbb{C}$, then $T \in \mathcal{B}_0(H)$ iff for every $\epsilon > 0$ there are only finitely many ν_γ of modulus at least ϵ.

3. Let $V \in \mathcal{B}(H)$ be such that V^2 is a compact positive operator. Is V necessarily compact? And if $V^2 = I$?

4. Let $S, T \in \mathcal{B}_0(H)$ be such that $\|Sx\| = \|Tx\|$ for all $x \in H$, with S a positive operator. Show that $S = (T^*T)^{1/2} = |T|$.

5. Let $T \in \mathcal{B}(\mathbb{C}^n)$ be a normal operator (i.e. $TT^* = T^*T$). Use elementary linear algebra to show that the matrix of T is diagonal in some orthonormal basis.

6. Let U be the compact operator defined by a hermitian kernel, as in Theorem 14. Show that if U is positive then $K(s,s) \geq 0$ for every s $(0 \leq s \leq 1)$. Show also that

$$K(s,t) - \sum_{k=1}^{n} \lambda_k \varphi_k(s)\overline{\varphi_k(t)}$$

is the kernel of a positive operator.
Deduce that

$$K(s,t) = \sum_{n} \lambda_n \varphi_n(s)\overline{\varphi_n(t)},$$

where the series is absolutely and uniformly convergent.

7. Let $(x_n)_1^\infty$ be an orthonormal basis consisting of eigenvectors of a compact hermitian operator U, with $Ux_n = \lambda_n x_n$. Let $\mu_1 \geq \mu_2 \geq \ldots$ be the multiset $\{\lambda_n : \lambda_n > 0\}$ arranged in a decreasing order, with $y_1, y_2, \ldots$ the corresponding eigenvectors. Putting it another way: let $\mu_1 \geq \mu_2 \geq \cdots > 0$ be the sequence of non-negative eigenvalues repeated according to their multiplicities. Show that

$$\mu_n = \max\{(Ux,x) : \|x\| = 1, x \perp y_i \text{ for } i = 1, \ldots, n-1\}.$$

Show also that

$$\mu_n = \min_{H_{n-1}} \max\{(Ux,x) : x \in H_{n-1}, \|x\| = 1\},$$

where the minimum is over all $(n-1)$-codimensional subspaces H_{n-1}.
Finally, show that

$$\mu_n = \max_{F_n} \min\{(Ux,x) : x \in H_n, \|x\| = 1\},$$

where the maximum is over all n-dimensional subspaces F_n.

8. Let U be a positive hermitian operator. Show that

$$\|Ux\|^4 \leq (Ux,x)(U^2x, Ux)$$

for every vector x. Deduce from this that $\|U\| = v(U)$.

9. Let $U \in \mathcal{B}_0(H)$ be a positive hermitian operator with $\text{Ker } U = \{0\}$. Show that there is a sequence of hermitian operators $(U_n)_1^\infty \subset \mathcal{B}(H)$ such that $U_n Ux \to x$ and $UU_nx \to x$ for every $x \in H$. Can one have $U_n U \to I$ as well?

10. Let $U \in \mathcal{B}_0(H)$ be hermitian. Prove that $\text{Im } U$ is a closed subspace of H iff U has finite rank.

11. Let $T \in \mathcal{B}_0(H)$. Prove that T is
    (a) normal iff H has an orthonormal basis consisting of eigenvectors of T;
    (b) hermitian iff it is normal and all its eigenvalues are real;
    (c) positive iff it is normal and all its eigenvalues are non-negative reals.

12. Let $U \in \mathcal{B}_0(H)$ be hermitian. Prove that there are unique positive operators $U_+, U_- \in \mathcal{B}_0(H)$ such that

$$U = U_+ - U_- \quad \text{and} \quad U_+ U_- = U_- U_+ = 0.$$

13. Prove the *Fredholm alternative for hermitian operators*: Let U be a compact hermitian operator on a Hilbert space H and consider the following two equations:

$$Ux - x = 0 \tag{2}$$

and

$$Ux - x = x_0 \tag{3}$$

where $x_0 \in H$. Then *either*
    (a) the only solution of (2) is $x = 0$, and then (3) has a unique solution,

or
    (b) there are non-zero solutions of (2), and then (3) has a solution iff x_0 is orthogonal to every solution of (2); furthermore, if (3) has a solution then it has infinitely many solutions: if x is a solution of (3) then x' is also a solution iff $x - x'$ is a solution of (2).

14. Give a detailed proof of Theorem 10. In particular, check that the map $\mathcal{C}_N(H) \to \mathcal{N}$ is a surjection.

15. Let $T \in \mathcal{B}_0(H)$ be normal and, as in Theorem 10, for $\lambda \in \mathbb{C}$ set $n_T(\lambda) = \dim \operatorname{Ker}(\lambda I - T)$. Prove that $n_T(\lambda) \leqslant 1$ for every $\lambda \in \mathbb{C}$ (including $\lambda = 0$) iff there is a *cyclic vector* for T, i.e. a vector $x_0 \in H$ such that $\operatorname{lin}\{x_0, Tx_0, T^2x_0, \ldots\}$ is dense in H.

Notes

There are many good accounts of applications of the spectral theorem for compact hermitian opertors to differential and integral equations. We followed J. Dieudonné, *Foundations of Modern Analysis*, Academic Press, New York and London, 1960, xiv + 361 pp. Here are some of the other good books to consult for the Sturm–Liouville problem, Green's functions, the use of the Fredholm alternative, etc: D. H. Griffel, *Applied Functional Analysis*, Ellis Horwood, Chichester, 1985, 390 pp., I. J. Maddox, *Elements of Functional Analysis*, 2nd edn., Cambridge University Press, 1988, xii + 242 pp., and N. Young, *An Introduction to Hilbert Space*, Cambridge University Press, 1988, vi + 239 pp.

15. FIXED-POINT THEOREMS

In Chapter 7 we proved the doyen of fixed-point theorems, the contraction-mapping theorem. In this chapter we shall prove some considerably more complicated results: Brouwer's fixed-point theorem and some of its consequences. It is customary to deduce Brouwer's theorem from some standard results in algebraic topology, but we shall present a self-contained combinatorial proof.

Before we can get down to work, we have to plough through some definitions.

A *flat* (or an *affine subspace*) of a vector space V is a set of the form $F = x + W$, where W is subspace of V. If W is k-dimensional then we call F a *k-flat*. As the intersection of a set of flats is either empty or a flat, for every set $S \subset V$ there is a minimal flat F containing S, called the *flat spanned by S*. Clearly

$$F = \left\{ \sum_{i=1}^{n} \lambda_i x_i : x_i \in S, \ \sum_{i=1}^{n} \lambda_i = 1, \ n = 1, 2, \ldots \right\}$$

Let $x_0, x_1, \ldots x_k$ be points in a vector space. We say that these points are *in general position* if the minimal flat containing them is k-dimensional, i.e. if the vectors $x_1 - x_0, x_3 - x_0, \ldots, x_k - x_0$ span a k-dimensional subspace. Equivalently, they are in general position if $\mu_0 = \mu_1 = \cdots = \mu_k = 0$ whenever $\sum_{i=0}^{k} \mu_i x_i = 0$ and $\sum_{i=0}^{k} \mu_i = 0$ or, in other words, if the points are distinct and $\{x_1 - x_0, x_2 - x_0, \ldots, x_k - x_0\}$ is a linearly independent set of vectors.

For $0 \leq k \leq n$, let $x_0, x_1, \ldots, x_k$ be $k + 1$ points in $\mathbb{R}^n$ in general position. The *k-simplex* $\sigma = (x_0, x_1, \ldots, x_k)$ with vertices $x_0, x_1, \ldots, x_k$ is the following subset of $\mathbb{R}^n$:

$$\left\{ \sum_{i=0}^{k} \mu_i x_i : \sum_{i=0}^{k} \mu_i = 1, \mu_i > 0 \text{ for all } i \right\}.$$

The *skeleton* of σ is the set $\{x_0, x_1, \ldots, x_k\}$ and the *dimension* of σ is k. Usually we write σ^k for a simplex of dimension k and call it a k-*simplex*. A 0-simplex is called a *vertex*.

A simplex σ_1 is a *face* of a simplex σ_2 if the skeleton of σ_1 is a subset of the skeleton of σ_2.

Note that the closure of the simplex $\sigma = (x_0, x_1, \ldots, x_k)$ in $\mathbb{R}^n$ is

$$\bar{\sigma} = [x_0, x_1, \ldots, x_k]$$

$$= \left\{ \sum_{i=0}^{k} \mu_i x_i : \sum_{i=0}^{k} \mu_i = 1, \mu_i \geqslant 0 \right\}$$

$$= \bigcup \{(x_{i_0}, x_{i_1}, \ldots, x_{i_r}) : \{i_0, i_1, \ldots, i_r\} \subset \{0, 1, \ldots, n\}\},$$

i.e. the closure of σ is precisely the union of all faces of σ, including itself. Also, $\bar{\sigma}$ is precisely $\mathrm{co}\{x_0, x_1, \ldots, x_k\}$, the convex hull of the vertices, and σ is the interior of this convex hull in the k-flat spanned by the vertices.

A finite set K of disjoint simplices in $\mathbb{R}^n$ is called a *simplicial complex* if every face of every simplex of K is also a simplex of K. We also call K a *simplicial decomposition* of the set $|K| = \bigcup \{\sigma : \dot{\sigma} \in K\}$, the *body* of K. If K is a simplicial complex and $\sigma, \tau \in K$ then the closed simplices $\bar{\sigma}$ and $\bar{\tau}$ are either disjoint or meet in a closed face of both.

We are ready to prove the combinatorial basis of Brouwer's theorem.

Lemma 1. (Sperner's lemma) Let K be a simplicial decomposition of a closed n-simplex $\bar{\sigma} = [x_0, x_1, \ldots, x_n]$. Let S be the set of vertices of K and let $\gamma : S \to \{0, 1, \ldots, n\}$ be an $(n+1)$-colouring of S such that the colours of the vertices contained in a face $[x_{i_0}, x_{i_1}, \ldots, x_{i_r}]$ of σ belong to $\{i_0, i_1, \ldots, i_r\}$. Call an n-simplex σ^n *multicoloured* if the vertices of σ^n are coloured with distinct colours. Then the number of multicoloured n-simplices of K is odd.

Proof. Let us apply induction on n. For $n = 0$ the assertion is trivial; so assume that $n \geqslant 1$ and the result holds for $n-1$.

Call an $(n-1)$-face of K *marked* if its vertices are coloured with $0, 1, \ldots, n-1$, with each colour appearing once. For an n-simplex $\sigma^n \in K$, denote by $m(\sigma^n)$ the number of marked $(n-1)$-faces of σ^n. Note that a multicoloured n-simplex has precisely one marked $(n-1)$-face, and an n-simplex, which is not multicoloured, has either no marked face or two marked faces. Therefore the theorem claims that

$$m(K) = \sum_{\sigma^n \in K} m(\sigma^n) \tag{1}$$

is odd.

Now let us look at the sum in (1) in another way. What is the contribution of an $(n-1)$-simplex $\sigma^{n-1} \in K$ to $m(K)$? If σ^{n-1} is not marked, the contribution is 0. In particular, if σ^{n-1} is in a closed $(n-1)$-face of σ other than $\bar{\sigma}_0 = [x_0, x_1, \ldots, x_{n-1}]$ then the contribution of σ^{n-1} is 0. If $\sigma^{n-1} \in K$ is in $\bar{\sigma}_0$ and is marked then the contribution of σ^{n-1} is 1 since σ^{n-1} is a face of exactly one n-simplex of K. Furthermore, if σ^{n-1} is in σ, i.e. in the interior of the original n-simplex, then σ^{n-1} contributes 1 to $m(\sigma^n)$ if σ^{n-1} is a face of σ^n: as there are two such n-simplices σ^n, the total contribution of σ^{n-1} to $m(K)$ is 2. Hence, modulo 2, $m(K)$ is congruent to the number of marked $(n-1)$-simplices in $\bar{\sigma}_0$. By the induction hypothesis, this number is odd. Therefore so is $m(K)$, completing the proof. $\qquad\square$

Given points $x_0, x_1, \ldots, x_k$ of $\mathbb{R}^n$ in general position, for every point x of the k-dimensional affine plane P_k through $x_0, x_1, \ldots, x_k$, there are unique reals $\lambda_1, \lambda_2, \ldots, \lambda_k$ such that

$$x = x_0 + \sum_{i=1}^{k} \lambda_i(x_i - x_0).$$

Hence, there are unique reals $\mu_0, \mu_1, \ldots, \mu_k$ such that $x = \sum_{i=0}^{k} \mu_i x_i$ and $\sum_{i=0}^{k} \mu_i = 1$. These μ_i $(i = 0, 1, \ldots, k)$ are called the *barycentric coordinates* of x with respect to $(x_0, x_1, \ldots, x_k)$. Also, if $\sum_{i=0}^{k} \mu_i = 1$ then $\sum_{i=0}^{k} \mu_i x_i \in P_k$. Furthermore, the closed half-space of P_k containing x_k and bounded by the $(k-1)$-flat spanned by $x_0, x_1, \ldots, x_{k-1}$ is characterized by $\mu_k \geq 0$.

The barycentric coordinates can be used to define a very useful simplicial decomposition. Given a simplicial complex K, the *barycentric subdivision* sd K of K is the simplicial decomposition of $|K|$ obtained as follows. For a simplex $\sigma = (x_0, x_1, \ldots, x_k) \in K$ set

$$c_\sigma = \frac{1}{k+1} \sum_{i=0}^{k} x_i;$$

thus c_σ is the *barycentre* of σ. The complex sd K consists of all simplices $(c_{\sigma_0}, c_{\sigma_1}, \ldots, c_{\sigma_k})$ such that $\sigma_i \in K$ and σ_i is a proper face of σ_{i+1} $(i = 0, 1, \ldots, k-1)$.

To define the r-times iterated barycentric subdivision of K, set $\mathrm{sd}^0 K = K$ and $\mathrm{sd}^r K = \mathrm{sd}(\mathrm{sd}^{r-1}K)$ for $r \geqslant 1$. Thus $\mathrm{sd}^1 K = \mathrm{sd}\, K$.

The *mesh* of K, written mesh K, is the maximal diameter of a simplex of K. Equivalently, it is the maximal length of a 1-simplex of K. Note that if $\sigma_i = (x_0, x_1, \ldots, x_i)$ $(i = 0, 1, \ldots, k)$ are faces of a k-simplex $\sigma = \sigma_k = (x_0, x_1, \ldots, x_k)$ and $\tau = (c_{\sigma_0}, c_{\sigma_1}, \ldots, c_{\sigma_k})$, then the diameter of τ is less than $k/(k+1)$ times the diameter of σ. Therefore, if K is any simplicial complex then for every $\epsilon > 0$ there is an r such that mesh $\mathrm{sd}^r K < \epsilon$.

Let Y be a subset of a topological space X, and let $\alpha = \{A_\gamma : \gamma \in \Gamma\}$ be a collection of subsets of X. We call α a *covering* of Y if $Y \subset \bigcup_{\gamma \in \Gamma} A_\gamma$. Furthermore, α is a *closed covering* if each A_γ is closed, and it is an *open covering* if each A_γ is open. In what follows, the underlying topological space X is always $\mathbb{R}^n$. Sperner's lemma has the following important consequence.

Corollary 2. Let $\{A_0, A_1, \ldots, A_n\}$ be a closed covering of a closed n-simplex $\bar{\sigma} = [x_0, x_1, \ldots, x_n]$ such that each closed face $[x_{i_0}, x_{i_1}, \ldots, x_{i_r}]$ of $\bar{\sigma}$ is contained in $\bigcup_{j=0}^{r} A_{i_j}$. Then $\bigcap_{i=0}^{n} A_i \neq \varnothing$.

Proof. As we may replace A_i by $A_i \cap \bar{\sigma}$, we may assume that each A_i is compact. The compactness of the sets $A_0, A_1, \ldots, A_n$ implies that it suffices to show that for every $\epsilon > 0$ there are points $a_i \in A_i$ $(i = 0, 1, \ldots, n)$ such that $|a_i - a_j| < \epsilon$ if $i \neq j$.

Let then $\epsilon > 0$. Let K be a triangulation of $\bar{\sigma}$ such that every simplex of K has diameter less than ϵ; as we have seen, for K we may take an iterated barycentric subdivision of $\bar{\sigma}$. Given a vertex x of K contained in a face $(x_{i_0}, x_{i_1}, \ldots, x_{i_r})$ of σ, we know that $x \in \bigcup_{j=0}^{r} A_{i_j}$. Set $\gamma(x) = \min\{i_j : x \in A_{i_j}\}$.

The colouring γ of the vertex set of K satisfies the conditions of Lemma 1 and so K has a multicoloured n-simplex

$$(a_0, a_1, \ldots, a_n) : \gamma(a_i) = i \qquad (i = 0, 1, \ldots, n).$$

But then $a_i \in A_i$, as required. $\qquad\qquad\qquad\qquad\qquad\square$

From here it is a short step to one of the most fundamental fixed-point theorems, namely Brouwer's fixed-point theorem. A *closed n-cell* is a topological space homeomorphic to a closed n-simplex.

Theorem 3. (Brouwer's fixed-point theorem) Every continuous mapping of a closed n-cell into itself has a fixed point.

Proof. We may assume that our n-cell is exactly a closed simplex $\bar{\sigma}^n = [x_0, x_1, \ldots, x_n]$. Suppose that $\varphi : \bar{\sigma}^n \to \bar{\sigma}^n$ is a continuous map, sending a point

$$x = \sum_{i=0}^{n} \mu_i x_i \qquad \left(\mu_i \geq 0, \ \sum_{i=0}^{n} \mu_i = 1 \right)$$

to

$$\varphi(x) = \sum_{i=0}^{n} \mu_i' x_i \qquad \left(\mu_i' \geq 0, \ \sum_{i=0}^{n} \mu_i' = 1 \right).$$

For each i, let

$$A_i = \{ x \in \sigma^n : \mu_i' \leq \mu_i \}.$$

Then $\{A_0, A_1, \ldots, A_n\}$ is a closed covering of $\bar{\sigma}^n$. If a point $x = \sum_{i=0}^{n} \mu_i x_i$ belongs to a closed face $[x_{i_0}, x_{i_1}, \ldots, x_{i_r}]$ of $\bar{\sigma}^n$ then $\mu_i = 0$ for $i \notin \{i_0, i_1, \ldots, i_r\}$ and so $\sum_{j=0}^{r} \mu_{i_j} = 1$. Since $\sum_{i=0}^{n} \mu_i' = 1$, there is an index j such that $\mu_{i_j}' \leq \mu_{i_j}$ and so $x \in A_{i_j}$. Consequently,

$$[x_{i_0}, x_{i_1}, \ldots, x_{i_r}] \subset \bigcup_{j=0}^{r} A_{i_j},$$

showing that the conditions of Corollary 2 are satisfied. Thus there is a point x in all the A_i; such an x is a fixed point of φ. $\qquad \square$

The following lemma enables us to apply Theorem 3 to a rather pleasant class of spaces, namely the compact convex subsets of finite-dimensional spaces, i.e. the bounded closed convex subsets of finite-dimensional spaces.

Lemma 4. Let K be a non-empty compact convex subset of a finite-dimensional normed space. Then K is an n-cell for some n.

Proof. We may assume that K contains at least two points (and hence it contains a segment) since otherwise there is nothing to prove.

We may also suppose that K is in a real normed space and hence that K is a compact convex subset of $l_2^n = (|R^n, \|\cdot\|)$ for some n. Furthermore, by replacing $\mathbb{R}^n$ by the flat spanned by K and translating it, if necessary, we may assume that $0 \in \text{Int } K$.

Finally, let $\bar{\sigma}$ be an n-simplex containing 0 in its interior, and define a homeomorphism $\varphi : K \to \bar{\sigma}$ as follows: for $x \in \mathbb{R}^n$ define

$$n(x) = n_K(x) = \inf\{t : t > 0, x \in tK\},$$

and

$$m(x) = m_{\bar{\sigma}}(x) = \inf\{t : t > 0, x \in t\bar{\sigma}\},$$

and for $x \in K$ set

$$\varphi(x) = \begin{cases} 0 & \text{if } x = 0, \\ \dfrac{n(x)}{m(x)} x & \text{if } x \neq 0. \end{cases} \qquad \square$$

Corollary 5. Let K be a non-empty compact convex subset of a finite-dimensional normed space. Then every continuous map $f : K \to K$ has a fixed point.

Proof. This is immediate from Theorem 3 and Lemma 4. $\square$

Our next aim is to prove an extension of Corollary 5 implying, in particular, that the corollary is true without the restriction that the normed space is finite-dimensional. This is based on the possibility of approximating a compact convex subset of a normed space by compact convex subsets of finite-dimensional subspaces. Unfortunately, the simple lemma we require needs a fair amount of preparation.

Let $S = \{x_1, \ldots, x_k\}$ be a finite subset of a normed space X. For $\epsilon > 0$ let $N(S, \epsilon)$ be the union of the open balls of radius ϵ centred at $x_1, \ldots, x_k$:

$$N(S, \epsilon) = \bigcup_{i=1}^{k} D(x_i, \epsilon).$$

For $x \in N(S, \epsilon)$ define $\lambda_i(x) = \max\{0, \epsilon - \|x - x_i\|\}$ $(i = 1, \ldots, k)$ and set $\lambda(x) = \sum_{i=1}^{k} \lambda_i(x)$. If $x \in N(S, \epsilon)$ then x belongs to at least one open ball $D(x_i, \epsilon)$, and for that index i we have $\lambda_i(x) > 0$. Hence $\lambda(x) > 0$ for every $x \in N(S, \epsilon)$. Define the *Schauder projection* $\varphi_{S, \epsilon} : N(S, \epsilon) \to \mathrm{co}\{x_1, \ldots, x_k\}$ by

$$\varphi_{S, \epsilon}(x) = \sum_{i=1}^{k} \frac{\lambda_i(x)}{\lambda(x)} x_i.$$

Here

$$\mathrm{co}\{x_1, \ldots, x_k\} = \left\{ \sum_{i=1}^{k} \lambda_i x_i : \lambda_i \geq 0, \sum_{i=1}^{k} \lambda_i = 1 \right\}$$

is the *convex hull* of the points $x_1, \ldots, x_k$: the intersection of all convex sets containing all the points $x - 1, \ldots, x_k$. This convex hull is, in fact, compact, since it is the continuous image of a closed $(k-1)$-simplex in $\mathbb{R}^k$. Indeed, if $(e_i)_1^k$ is the standard basis of $l_2^k = (\mathbb{R}^k, \|\cdot\|)$, say, then the closed simplex $\bar{\sigma} = [e_1, \ldots, e_k]$ is a bounded closed subset of l_2^k, and so it is compact. Furthermore, $\varphi : \bar{\sigma} \to \mathrm{co}\{x_1, \ldots, x_k\}$, given by

$$\sum_{i=1}^{k} \lambda_i e_i \mapsto \sum_{i=1}^{k} \lambda_i x_i, \qquad \text{where } \lambda_i \geq 0 \text{ and } \sum_{i=1}^{k} \lambda_i = 1,$$

is a continuous map.

Lemma 6. The Schauder projection $\varphi_{S,\epsilon}$ is a continuous map from $N(S, \epsilon)$ to $\mathrm{co}\{x_1, \ldots, x_n\}$ and

$$\|\varphi_{S,\epsilon}(x) - x\| < \epsilon \tag{2}$$

for all $x \in N(S, \epsilon)$.

Proof. Only (2) needs any justification. If $x \in N(S, \epsilon)$ then

$$\varphi_{S,\epsilon}(x) - x = \sum_{i=1}^{k} \frac{\lambda_i(x)}{\lambda(x)}(x_i - x) = \sum_{\lambda_i(x)>0} \frac{\lambda_i(x)}{\lambda(x)}(x_i - x).$$

But if $\lambda_i(x) > 0$ then $\|x_i - x\| < \epsilon$ and so

$$\|\varphi_{S,\epsilon}(x) - x\| \leq \sum_{\lambda_i(x)>0} \frac{\lambda_i(x)}{\lambda(x)}\|x_i - x\| < \epsilon. \qquad \square$$

Here then is the promised extension of Corollary 5, *Schauder's fixed-point theorem*.

Theorem 7. Let A be a (non-empty) closed convex subset of a normed space X and let $f : A \to A$ be a continuous map such that $K = \overline{f(A)}$ is compact. Then f has a fixed point.

Proof. Let $n \geq 1$. As K is compact, there is a finite set

$$S_n = \{x_1, \ldots, x_{k_n}\} \subset K$$

such that

$$K \subset \bigcup_{i=1}^{k_n} D\left(x_i, \frac{1}{n}\right) = N\left(S_n, \frac{1}{n}\right).$$

Set $K_n = \mathrm{co}\{x_1, \ldots, x_{k_n}\}$ and denote by φ_n the Schauder projection $\varphi_{S_n, 1/n} : N(S_n, 1/n) \to K_n$. We have $K_n \subset A$, so by Lemma 6 the restriction of $\varphi_n \circ f$ to K_n is a continuous map of K_n into itself. Hence, by Corollary 5, there is a point $x_n \in K_n$ such that $\varphi_n(f(x_n)) = x_n$. Therefore, by (2),

$$\|f(x_n) - x_n\| < \frac{1}{n} . \tag{3}$$

As each $f(x_n)$ belongs to the compact set K, the sequence $(f(x_n))_1^\infty$ has a convergent subsequence, say $f(x_{n_k}) \to x$ as $k \to \infty$, where $x \in K$. But then, by (3), $x_{n_k} \to x$ as $k \to \infty$, and so $f(x) = x$. $\square$

As a beautiful application of Brouwer's theorem, we prove Perron's theorem concerning eigenvalues of positive matrices.

Theorem 8. A matrix whose entries are all positive has a positive eigenvalue with an eigenvector whose coordinates are all positive.

Proof. Let $A = (a_{ij})$ be an $n \times n$ matrix with $a_{ij} > 0$ for all i and j. Let $(e_i)_1^n$ be the standard basis in $\mathbb{R}^n$. The closed $(n-1)$-simplex $\bar{\sigma} = [e_1, \dots, e_n]$ is a 'face' of the unit sphere

$$S(l_1^n) = \left\{ x = (x_i)_1^n \in \mathbb{R}^n : \|x\|_1 = \sum_{i=1}^n |x_i| = 1 \right\}.$$

The continuous map $\bar{\sigma} \to \bar{\sigma}$ given by $x \mapsto Ax/\|Ax\|_1$ has a fixed point $x = (x_i)_1^n$. Clearly, $Ax = \lambda x$ for some $\lambda > 0$ and $x_i > 0$ for all i. $\square$

From Theorem 7 it is a short step to a version of the Markov–Kakutani fixed-point theorem. An *affine map* of a vector space V into itself is a map of the form $x \mapsto x_0 + S(x)$, where $S : V \to V$ is a linear map. Equivalently, $T : V \to V$ is an affine map if

$$T\left(\sum_{i=1}^n \lambda_i x_i \right) = \sum_{i=1}^n \lambda_i T(x_i)$$

whenever $x_i \in V$, $\lambda_i \geq 0$ and $\sum_{i=1}^n \lambda_i = 1$.

Theorem 9. Let K be an non-empty compact convex subset of a normed space X and let $\mathcal{F}$ be a commuting family of continuous affine maps on X such that $T(K) \subset K$ for all $T \in \mathcal{F}$. Then some $x_0 \in K$ is a fixed point of all the maps $T \in \mathcal{F}$.

Proof. For $T \in \mathcal{F}$ let K_T be the set of fixed points of T in K:

$$K_T = \{ x \in K : Tx = x \}.$$

By Theorem 7, $K_T \neq \emptyset$ and, as T is a continuous affine map, K_T is a compact convex subset of K. If $S \in \mathcal{F}$ then S maps K_T into itself since if $Tx = x$ then $T(Sx) = S(Tx) = Sx$ and so $Sx \in K_T$. Consequently, if for some $T_1, \dots, T_n \in \mathcal{F}$ and $S \in \mathcal{F}$

$$\bigcap_{i=1}^{n} K_{T_i} \neq \varnothing,$$

then $\bigcap_{i=1}^{n} K_{T_i}$ is a compact convex set mapped into itself by S. Hence, by Theorem 7,

$$K_S \cap \bigcap_{i=1}^{n} K_{T_i} \neq \varnothing.$$

This implies that the family of sets $\{K_T : T \in \mathcal{F}\}$ has the finite-intersection property. As each K_T is compact, there is a point x_0 which belongs to every K_T, i.e. $Tx_0 = x_0$ for every $T \in \mathcal{F}$. $\qquad \square$

One should remark that it is easy to prove Theorem 9 without relying on Theorem 7. Indeed, for $T \in \mathcal{F}$ and $n \geq 1$ the affine map

$$T^{(n)} = \frac{1}{n} \sum_{i=1}^{n} T^i$$

maps K into itself and $\mathcal{F}^* = \{T^{(n)} : T \in \mathcal{F}, n \geq 1\}$ is a commuting family of affine maps of K into itself. From this it follows that the system of compact sets $\{S(K) : S \in \mathcal{F}^*\}$ has the finite-intersection property. Hence there is a point x_0 such that $x_0 \in T^{(n)}(K)$ for every $T \in \mathcal{F}$ and $n \geq 1$.

This point x_0 is a fixed point of every $T \in \mathcal{F}$. Indeed, if $T^{(n)}(y_n) = x_0$ for $y_n \in K$ then

$$T(x_0) - x_0 = \frac{1}{n} \{T^n(y_n) - y_n\}.$$

Since $(y_n)_1^\infty$ is a bounded sequence, we have $T(x_0) = x_0$.

Exercises

1. Let X be a Banach space and let $f : B(X) \to X$ be a contraction from the closed unit ball into X (i.e. $d(f(x), f(y)) \leq k d(x, y)$ for all $x, y \in B(X)$ and some $k < 1$). By considering the map $g(x) = \frac{1}{2}\{x + f(x)\}$, or otherwise, prove that if $f(S(X)) \subset B(X)$ then f has a fixed point.

2. Deduce the following assertion from Corollary 2.

   Let $\{A_0, A_1, \ldots, A_n\}$ be a closed covering of a closed simplex $\bar{\sigma} = [x_0, x_1, \ldots, x_n]$ such that, for each i $(0 \leq i \leq n)$ the set A_i is disjoint from the closed $(n-1)$-face $\bar{\sigma}_i^{(n-1)}$ not containing x_i (i.e. 'opposite' the vertex x_i). Then $\bigcap_{i=0}^{n} A_i \neq \varnothing$.

3. Use the result in the previous exercise to prove that if $\bar{\sigma} = [x_0, x_1, \ldots, x_n]$ is a closed simplex and $f: \bar{\sigma} \to \bar{\sigma}$ is a continuous map such that for every closed $(k-1)$-face $\bar{\tau}$ of $\bar{\sigma}$ we have $f(\bar{\tau}) \subset \bar{\tau}$, then f is a surjection.

4. Prove that Brouwer's fixed-point theorem is equivalent to each of the following three assertions, where $B^n = B(l_1^n)$ and $S^{n-1} = S(l_1^n)$. (In fact, we could take $B^n = B(X)$ and $S^{n-1} = S(X)$ for any n-dimensional real normed space X.)

   (i) S^{n-1} is *not contractible in itself*, i.e. there is no continuous map $\Phi: S^{n-1} \times [0,1] \to S^{n-1}$ such that for some $x_0 \in S^{n-1}$ we have $\Phi(x, 0) = x_0$ and $\Phi(x, 1) = x$ for all $x \in S^{n-1}$.

   (ii) There is *no retraction* from B^n onto S^{n-1}, i.e. there is no continuous map $f: B^n \to S^{n-1}$ such that $f(x) = x$ for all $x \in S^{n-1}$.

   (iii) Whenever $f: B^n \to \mathbb{R}^n$ is a continuous map without a fixed point then there is a point $x \in S^{n-1}$ such that $x = \lambda f(x)$ for some $0 < \lambda < 1$.

5. Let C be a closed convex subset of a Hilbert space H. Show that for every $x \in H$ there is a unique point of C nearest to x, i.e. there is a unique point $\varphi(x) \in C$ such that

$$d(x, \varphi(x)) = \inf\{d(x, y): y \in C\}.$$

Show also that the function $\varphi: H \to C$ is continuous.

6. Combine Theorem 3 with the assertion in the previous exercise to deduce Corollary 5.

7. Let $C_n = B(l_\infty^n)$ be the n-dimensional cube:

$$C_n = \{x = (x_i)_1^n \in \mathbb{R}^n: |x_i| \leqslant 1 \text{ for every } i\}.$$

The *closed faces* of C_n are

$$F_i^+ = \{x \in C_n: x_i = 1\} \quad \text{and} \quad F_i^- = \{x \in C_n: x_i = -1\}$$

$(i = 1, \ldots, n)$. For each i $(i = 1, \ldots, n)$ let A_i be a closed subset of C_n *separating* F_i^+ and F_i^-, i.e. $C_n \backslash A_i = U_i^+ \cup U_i^-$, where U_i^+ and U_i^- are disjoint open subsets of C_n, with $F_i^+ \subset U_i^+$ and $F_i^- \subset U_i^-$. Prove that $\bigcap_{i=1}^n A_i \neq \varnothing$.

8. Let φ be a continuous one-to-one map from a compact Hausdorff space K into a Hausdorff space. Show that φ is a homeomorphism between K and $\varphi(K)$.

9. Prove that every compact metric space is homeomorphic to a closed subset of the *Hilbert cube*:

$$I^\infty = \{(x_i)_1^\infty \in l_2: |x_i| \leqslant 2^{-i} \text{ for every } i\}.$$

10. Prove that every continuous map of the Hilbert cube I^∞ into itself has a fixed point. (Check that the map $f: B(l_2) \to B(l_2)$ given by $f(x) = Sx + (1 - \|x\|)e_1$ has no fixed point, where S is the right shift and $e_1 = (1, 0, 0, \ldots)$.)

11. Show that a continuous map of $B(l_2)$ into itself need not have a fixed point.

12. Let C be a closed subset of a compact metric space K, and let f be a continuous map of C into a normed space X. Use Schauder projections and the Tietze–Urysohn extension theorem (Theorem 6.3) to prove that f has a continuous extension $F: K \to X$.

The aim of the next three exercises is to make it easy for the reader to prove another beautiful fixed-point theorem.

13. Let a, b, c and x be points in a Hilbert space such that $b = \frac{1}{2}(a + c)$ and

$$0 < r \leq \|x - a\| \leq \|x - b\| \leq \|x - c\| \leq r + \epsilon \leq 2r.$$

Deduce from the parallelogram law that

$$\|a - c\| \leq 4\delta,$$

where $8\delta^2 = 2r\epsilon + \epsilon^2$, and so $\|a - c\| \leq 2\sqrt{2r\epsilon}$.

14. Let C be a subset of $B(l_2)$, and let $f: C \to C$ be a *non-expansive map*, i.e. let f be such that $d(f(x), f(y)) \leq d(x, y)$ for all $x, y \in C$. Suppose x_1, x_2 and $a = \frac{1}{2}(x_1 + x_2) \in C$, and $\|f(x_i) - x_i\| \leq \epsilon$ for $i = 1, 2$. Set $c = f(a)$ and $b = \frac{1}{2}(a + c)$. Assuming that

$$\|x_1 - b\| \leq \|x_2 - b\|,$$

check that

$$\|x_1 - b\| \geq \|x_1 - a\|$$

and

$$\|x_1 - c\| \leq \|x_1 - a\| + \epsilon.$$

Deduce from the result in the previous exercise that

$$\|a - f(a)\| \leq 2\sqrt{\epsilon}.$$

15. Let C be a non-empty closed convex subset of $B(l_2)$, and let $f: C \to C$ be a non-expansive map. Set

$$F_n = \{x \in C : \|f(x) - x\| \leq 1/n\},$$

and show that $F_n \neq \varnothing$ for all n.

Put $d_n = \inf\{\|x\| : x \in F_n\}$ and note that the monotone increasing sequence $(d_n)_1^\infty$ converges to some $d \leq 1$. Use the result of the previous exercise to show that

$$\operatorname{diam} F_n = \sup\{\|x - y\| : x, y \in f_n\} \to 0 \qquad \text{as } n \to \infty.$$

Conclude from this that f has a fixed point.

16. Let $K(s, t)$ be continuous for $0 \leq s, t \leq 1$, and let $f(t, u)$ be continuous and bounded for $0 \leq t \leq 1$ and $-\infty < u < \infty$. Suppose that $|K(s, t)| \leq M$ and $|f(t, u)| \leq N$ for all s, t and u $(0 \leq s, t \leq 1)$. Define an operator $T : C[0, 1] \to C[0, 1]$ by

$$(Tu)(s) = \int_0^1 K(s, t) f(t, u(t)) \, dt.$$

Check that T maps $C[0, 1]$ into the closed ball of radius MN, say $C = \{u \in C[0, 1] : \|u\| \leq MN\}$.

Apply the Arzelà–Ascoli theorem to show that TC is a relatively compact subset of $C[0, 1]$.

Make use of the Schauder fixed-point theorem to prove that the *Hammerstein equation*

$$u(s) = \int_0^1 K(s, t) f(t, u(t)) \, dt$$

has a continuous solution.

Notes

This chapter is based on the book of J. Dugundji and A. Granas, *Fixed Point Theory*, vol. I, Polish Scientific Publishers, Warsaw, 1982, 209 pp., which is a rich compendium of beautiful results. Another interesting book on the topic is D. R. Smart, *Fixed Point Theorems*, Cambridge University Press, 1974. The more usual approach to fixed-point theorems is via homology, homotopy and degrees of maps; this can be found in most books on algebraic topology. The case $n = 3$ of Brouwer's fixed-point theorem (Theorem 3) was proved in L. E. J. Brouwer, *On continuous one-to-one transformations of surfaces into themselves*, Proc. Kon. Ned. Ak. V. Wet. Ser. A, **11** (1909), 788–98; the first proof of the full result was given by J. Hadamard, *Sur quelques* *applications de l'indice de Kronecker*; Appendix in J. Tannery, *Introduction à la Théorie des Fonctions d'une Variable*, vol. II, 2me éd., 1910; Brouwer himself proved the general case in 1912. Sperner's lemma is from E. Sperner, *Neuer Beweis für die Invarianz der Dimensionzahl und des Gebietes*, Abh. Math. Sem. Hamb. Univ., **6** (1928), 265–72, and Theorem 7 is from J. Schauder, *Der Fixpunktsatz in Funktionalräumen*, Studia Math., **2** (1930), 171–80. The original version of Theorem 9 is in A. A. Markoff, *Quelques théorèmes sur les ensembles abéliens*, C. R. Acad. Sci. URSS (N.S), **1** (1936), 311–3.

16. INVARIANT SUBSPACES

Given a complex Banach space X, which operators $T \in \mathcal{B}(X)$ have non-trivial closed invariant subspaces? This question, the so-called *invariant-subspace problem*, is the topic of this brief last chapter. Until fairly recently, it was not known whether there was any operator T without a non-trivial (closed) invariant subspace, and it is still not known whether there is such an operator on a (complex) Hilbert space.

Much of the effort concerning the invariant-subspace problem has gone into proving positive results, i.e. results claiming the existence of invariant subspaces for operators satisfying certain conditions. Our main aim in this chapter is to present the most beautiful of these positive results, Lomonosov's theorem, whose proof is surprisingly simple.

As we remarked earlier, the Riesz theory of compact operators on Banach spaces culminated in a very pleasing theorem, Theorem 13.8, which nevertheless, did not even guarantee the existence of a *single* non-trivial invariant subspace. This deficiency was put right, with plenty to spare, in Chapter 14, but only for a compact *normal* operator on a Hilbert space. Now we return to the general case to prove Lomonosov's theorem, which claims considerably more than that every compact operator has a non-trivial invariant subspace. Before we present this result, we need some definitions and a basic result about compact convex sets.

As in Chapters 13 and 14, all spaces considered in this chapter are *complex* spaces. Furthermore, as every linear operator on a finite-dimensional complex vector space has an eigenvector, we shall consider only *infinite-dimensional* spaces.

Given a Banach space X and an operator $T \in \mathcal{B}(X)$, call a subspace $Y \subset X$ *invariant under* T or T-*invariant* if Y is closed and $TY \subset Y$. We also say that Y is an *invariant subspace of T*. Clearly, $Y = \{0\}$ and $Y = X$ are T-invariant subspaces for every T, so we are interested only in other invariant subspaces, the so-called *non-trivial* invariant subspaces.

We call a subspace Y a *hyperinvariant subspace for T* if Y is an S-invariant subspace for every $S \in \mathscr{B}(X)$ commuting with T. Since T commutes with itself, every hyperinvariant subspace is an invariant subspace.

Note that if T and S commute then $\operatorname{Ker} T$ is S-invariant since if $x \in \operatorname{Ker} T$ then $TS(x) = ST(x) = S(0) = 0$ and so $S(x) \in \operatorname{Ker} T$. Hence if T has no non-trivial hyperinvariant subspaces then either T is a multiple of the identity or it has no eigenvalue, i.e. $\sigma_p(T) = \varnothing$. In particular, if $T \in \mathscr{B}_0(X)$ and $\sigma(T) \neq \{0\}$ then, by Theorem 8 of Chapter 13, the operator T has a non-trivial hyperinvariant subspace.

What is the easiest way of constructing a T-invariant subspace? Pick a vector $x \in X$ ($x \neq 0$) and set

$$Y_0(x) = \operatorname{lin}\{x, Tx, T^2x, \ldots\}.$$

Then $TY_0(x) \subset Y_0(x)$ and so, as T is continuous,

$$Y(x) = \overline{Y_0(x)} = \overline{\operatorname{lin}}\{x, Tx, T^2x, \ldots\}$$

is a T-invariant subspace. Hence if T has no non-trivial invariant subspace then $Y(x) = X$ for all $x \neq 0$. A vector x is said to be a *cyclic vector for T* if $Y(x) = X$.

Thus if T has no non-trivial invariant subspace then every non-zero vector is a cyclic vector for T. The converse of this is also trivially true: if every non-zero vector is a cyclic vector for T then T has no non-trivial invariant subspace since if Y is a non-trivial invariant subspace then no vector in Y is a cyclic vector for T.

Unfortunately, but not surprisingly, this shallow argument is no advance on the invariant-subspace problem; nevertheless, it tells us that we have to concentrate on cyclic vectors.

In the proof of Lomonosov's theorem, we need a basic result concerning closed convex hulls of compact sets.

Theorem 1. (Mazur's theorem) The closed convex hull of a compact set in a Banach space is compact.

Proof. Let A be a compact subset of a Banach space X and let $K = \overline{\operatorname{co}}\,A$. We have to show that K is totally bounded, i.e. for every $\epsilon > 0$ it contains a finite ϵ-net. Since A is compact, it contains a finite

$\frac{1}{3}\epsilon$-net, say $\{x_1,\ldots,x_n\}$. Thus $\{x_1,\ldots,x_n\} \subset A$ and for every $x \in A$ there is an x_i such that $\|x-x_i\| < \frac{1}{3}\epsilon$. The set $P = \mathrm{co}\{x_1,\ldots,x_n\}$ is also compact, and so it contains a finite $\frac{1}{3}\epsilon$-net $\{y_1,\ldots,y_m\}$. Furthermore, the set

$$P_{\frac{1}{3}\epsilon} = \{x \in X : d(x,P) < \tfrac{1}{3}\epsilon\}$$

is convex and contains A, and therefore contains $\mathrm{co}\,A$. But then with

$$M = \{x \in X : \|x-y_i\| < \tfrac{2}{3}\epsilon \text{ for some } i\}$$

we have $\mathrm{co}\,A \subset P_{\frac{1}{3}\epsilon} \subset M$ and so $\{y_1,\ldots,y_m\}$ is an ϵ-net in $K = \overline{\mathrm{co}}\,A$. $\qquad\square$

Here then is the main result of the Chapter.

Theorem 2. (Lomonosov's first theorem) Let T be a non-trivial compact operator on an infinite-dimensional complex Banach space X. Then T has a non-trivial hyperinvariant subspace.

Proof. We may assume that $\|T\| = 1$. Set

$$SA = \{T\}' = \{S \in \mathcal{B}(X) : ST = TS\}$$

and pick a point $x_0 \in X$ such that $\|x_0\| > \|T\|$. Set $B_0 = B(x_0, 1)$ and note that $0 \notin B_0$ and $0 \in \overline{TB_0}$.

Suppose first that there is a point $y_0 \in X$ ($y_0 \neq 0$) such that

$$\|T'y_0 - x_0\| \geqslant 1 \tag{1}$$

for every $T' \in SA$. Then we are done since

$$Y = \overline{\mathrm{lin}}\{T'y_0 : T' \in SA\}$$

is a non-trivial hyperinvariant subspace of X.

Suppose then that there is no $y_0 \neq 0$ satisfying (1). Then for every $y \in X$ ($y \neq 0$) there is an operator $T' \in SA$ such that

$$\|T'y - x_0\| < 1.$$

Since $\overline{TB_0}$ is a compact set not containing 0, there are operators $T_1,\ldots,T_n \in SA$ such that for every $y \in \overline{TB_0}$ there is a T_i ($1 \leqslant i \leqslant n$) satisfying

$$\|T_i y - x_0\| < 1. \tag{2}$$

Now we define a map $\psi : \overline{TB_0} \to X$ reminiscent of the Schauder projection. For $y \in \overline{TB_0}$ and $1 \leqslant i \leqslant n$ set

$$\lambda_i(y) = \max\{0, 1 - \|T_i y\|\}$$

and

$$\lambda(y) = \sum_{i=1}^{n} \lambda_i(y).$$

Relation (2) implies that $\lambda(y) > 0$ for every $y \in \overline{TB_0}$. Therefore we may define

$$\psi(y) = \sum_{i=1}^{n} \frac{\lambda_i(y)}{\lambda(y)} T_i y.$$

This map $\psi : \overline{TB_0} \to X$ is continuous and so $\overline{TB_0}$ is a compact subset of B_0. Consequently, by Mazur's theorem, $K = \overline{\text{co}}\, \overline{TB_0}$ is a compact convex subset of B_0. Hence

$$\psi \circ T : K \to K$$

is a continuous map of a compact convex set into itself and so, by Schauder's theorem (Theorem 15.7), it has a fixed point $z_0 \in K$:

$$\sum_{i=1}^{n} \frac{\lambda_i(Tz_0)}{\lambda(Tz_0)} T_i Tz_0 = z_0.$$

Set

$$S = \sum_{i=1}^{n} \frac{\lambda_i(Tz_0)}{\lambda(Tz_0)} T_i T.$$

Then $S \in SA$, $Sz_0 = z_0 \neq 0$, and so

$$Y = \text{Ker}(I - S) \neq \{0\}$$

is a T-invariant subspace. As S is compact, Y is finite-dimensional. But then $T|Y$ is an operator on a complex finite-dimensional space and so it has an eigenvalue λ. However then $\text{Ker}(\lambda I - T)$ is a non-trivial hyperinvariant subspace for T. □

A slight variation in the proof shows that an even larger class of operators have hyperinvariant subspaces.

Theorem 3. (Lomonosov's second theorem) If $T \in \mathcal{B}(X)$ commutes with a non-zero compact operator and is not a multiple of the identity then it has a hyperinvariant subspace.

Proof. Suppose $TT_0 = T_0T$ for some $T_0 \in \mathscr{B}_0(X)$ ($T_0 \neq 0$). Proceed as in the proof of Theorem 2 but consider $\overline{T_0B_0}$ instead of $\overline{TB_0}$ and so put $K = \overline{\mathrm{co}}\,\overline{T_0B_0}$. Then we obtain a fixed point $z_0 \in K$ of $\psi \circ T_0 : K \to K$, i.e.

$$Sz_0 = z_0$$

for

$$S = \sum_{i=1}^{n} \frac{\lambda_i(T_0 z_0)}{\lambda(T_0 z_0)} T_i T_0.$$

Therefore

$$Y = \mathrm{Ker}(I - S) \neq \{0\}$$

is a finite-dimensional T-invariant subspace, and now the proof is completed as before.   □

Let us point out the following special case of Theorem 3.

Corollary 4. Let $S, T \in \mathscr{B}(X)$ be commuting operators such that S commutes with a non-zero compact operator, and is not a multiple of the identity. Then T has a non-trivial invariant subspace.   □

As a rather special case of this corollary one obtains a theorem of Aronszajn and Smith, proved considerably earlier.

Corollary 5. Every compact operator on an infinite-dimensional complex Banach space has a non-trivial invariant subspace.   □

An extension of this result, first proved by Bernstein and Robinson, needs only a little work.

Corollary 6. Let X be an infinite-dimensional complex Banach space and let $T \in \mathscr{B}(X)$ be such that $p(T) \in \mathscr{B}_0(X)$ for some non-zero complex polynomial $p(z)$. Then T has a non-trivial invariant subspace.

Proof. Let

$$p(z) = \sum_{k=0}^{n} a_k z^k, \qquad \text{with} \, a_n \neq 0.$$

If $p(T) \neq 0$ then, as $Tp(T) = p(T)T$, the assertion follows from Theorem 2.

If, on the other hand, $p(T) = 0$, then $a_n T^n = -\sum_{k=0}^{n-1} a_k T^k$, and so $Y(x) = \text{lin}\{x, Tx, \ldots, T^{n-1}x\}$ is a T-invariant subspace for every $x \neq 0$.

$\square$

In spite of the simplicity of its proof, Lomonosov's second theorem is a very powerful result. At the moment, it is not clear how large a class of operators Corollary 4 applies to; in fact, for a while it was not clear that there is *any* operator $T \in \mathcal{B}(X)$ which is not covered by Corollary 4.

The invariant-subspace problem for Banach spaces was solved, in the negative, only fairly recently: Per Enflo and Charles Read constructed complex Banach spaces and bounded linear operators on them which do not have non-trivial invariant subspaces. The original proofs were formidably difficult and the spaces seemed to be rather peculiar spaces. Later, Charles Read gave an easily accessible proof, and showed that his construction works, in fact, on l_1.

In view of these great results, the invariant-subspace problem for Hilbert spaces has become a very major problem in functional analysis. In fact, it is not impossible that the answer is in the affirmative even on reflexive spaces, i.e. that every bounded linear operator on an infinite-dimensional reflexive complex Banach space has a non-trivial invariant subspace.

Exercises

1. Let X be a non-separable Banach space. Show that every $T \in \mathcal{B}(X)$ has a non-trivial invariant subspace.
2. Show that the following result can be read out of the proof of Theorem 2. Let SA be a subalgebra of $\mathcal{B}(X)$ whose elements do not have a non-trivial common invariant subspace. Then if $T \in \mathcal{B}_0(X)$ and $T \neq 0$ then there is an operator $A \in SA$ such that $\text{Ker}(I - AT) \neq \{0\}$.
3. Let $T_1, \ldots, T_n \in \mathcal{B}_0(X)$ be commuting operators. Show that they have a non-trivial common invariant subspace.
4. Deduce from Theorem 1 and Exercise 5.5 the following extension of Theorem 4.10. If the unit ball of a *Banach* space X is σ-compact then X is finite-dimensional.

5^{+++}. Solve the invariant-subspace problem for Hilbert spaces.

Notes

Mazur's theorem is from S. Mazur, *Über die kleinste konvexe Menge, die eine gegebene kompakte Menge enthält*, Studia Math., **2** (1930), 7–9. Theorems 2 and 3 are from V. I. Lomonosov, *On invariant subspaces of families of operators, commuting with a compact operator* (in Russian), Funk. Analiz i ego Prilozh, **7** (1973), 55–6; to be precise, Theorem 3 is given as a remark added in proof. Corollary 5 is from N. Aronszajn and K. Smith, *Invariant subspaces of completely continuous operators*, Ann. Math., **60** (1954), 345–50.

The invariant-subspace problem for Banach spaces was solved in P. Enflo, *On the invariant subspace problem in Banach spaces*, Acta Math., **158** (1987), 213–313, and C. J. Read, *A solution to the Invariant Subspace Problem*, Bull. London Math. Soc., **16** (1984), 337–401.

A simplified and stronger version of Enflo's solution can be found in B. Beauzamy, *Un opérateur sans sous-espace invariant non-trivial: simplification de l'example de P. Enflo*, Integral Equations and Operator Theory, **8** (1985), 314–84.

Read's result concerning l_1 is in *A solution to the Invariant Subspace Problem on the space l_1*, Bull. London Math. Soc., **17** (1985), 305–17.

An interesting account of the results concerning the invariant-subspace problem can be found in B. Beauzamy, *Introduction to Operator Theory and Invariant Subspaces*, North Holland, Amsterdam, 1988, xiv + 358 pp.

INDEX OF NOTATION

INDEX OF TERMS

absolute value of an operator, 205
absolutely convergent series, 36
absolutely convex set, 27
adjoint of an operator, 155
adjoint operator, 31
affine hyperplane, 46
affine map, 220
affine subspace, 213
Alaoglu's theorem, 118
algebra, commutative, 92
algebraic dual of a normed space, 45
AM-GM inequality, 1
analytic, 171
annihilator of a set, 164
annihilator of a subspace, 158
approximate eigenvector, 169
approximate point spectrum, 169
approximation problem, 189
arithmetic mean, 1, 6
  weighted, 7
Arzelà–Ascoli theorem, 90
Auerbach system, 65

Banach algebra, 92
  unital, 32
Banach limit, 59
Banach space, 21

Banach's fixed-point theorem, 101
Banach–Mazur distance, 66
Banach–Steinhaus theorem, 78
barycentre, 215
barycentric coordinates, 215
barycentric subdivision, 215
basic sequence, 72
basis, 19, 83
  canonical, 37
  Hamel, 42
  Schauder, 37, 83
  standard, 37
basis constant, 73, 83
Bernstein and Robinson, theorem of,
  230
Bessel's inequality, 147
biorthogonal system, 64
  normalised, 64
Bishop–Phelps–Bollobás theorem, 122
body, 214
bounded below, 162
bounded linear operator, 28
bracket notation, 28
Brouwer's fixed-point theorem, 216

canonical basis, 37
Carleson's theorem, 150

Printed in the United States
By Bookmasters